HEAD-ORDER TECHNIQUES AND OTHER PRAGMATICS OF LAMBDA CALCULUS GRAPH REDUCTION

Nikos B. Troullinos

DISSERTATION.COM

Boca Raton

Head-Order Techniques and Other Pragmatics of Lambda Calculus Graph Reduction

Dissertation.com
Boca Raton, Florida
USA • 2011

ISBN-10: 1-61233-757-0
ISBN-13: 978-1-61233-757-9

Cover image © Michael Travers/Cutcaster.com

Abstract

In this dissertation Lambda Calculus reduction is studied as a means of improving the support for declarative computing. We consider systems having reduction semantics; i.e., systems in which computations consist of equivalence-preserving transformations between expressions. The approach becomes possible by reducing expressions beyond weak normal form, allowing expression-level output values, and avoiding compilation-centered transformations. In particular, we develop reduction algorithms which, although not optimal, are highly efficient.

A minimal linear notation for lambda expressions and for certain runtime structures is introduced for explaining operational features. This notation is related to recent theories which formalize the notion of substitution.

Our main reduction technique is Berkling's Head Order Reduction (HOR), a delayed substitution algorithm which emphasizes the extended left spine. HOR uses the de Bruijn representation for variables and a mechanism for artificially binding relatively free variables. HOR produces a lazy variant of the head normal form, the natural midway point of reduction. It is shown that beta reduction in the scope of relative free variables is not hard. Full normalization suggests new applications by not relegating partial evaluation to a meta level. Variations of HOR are presented, including a conservative "breadth-first" one which takes advantage of the inherent parallelism of the head normal form.

A reduction system must be capable of sharing intermediate results. Sharing under HOR has not received attention to date. In this dissertation variations of HOR which achieve sharing are described. Sharing is made possible via the special treatment of expressions referred to by head variables. The reduction strategy is based on normal order, achieves low reduction counts, but is shown to be incomplete.

Head Order Reduction with and without sharing, as well as other competing algorithms are evaluated on several test sets. Our results indicate that reduction rates in excess of one million reductions/second can be achieved on current processors in interpretive mode and with minimal pre- and post-processing. By extending the efficient algorithms for the pure calculus presented in this dissertation with primitives and data structures it is now possible to build useful reduction systems. We present some suggestions on how such systems can be designed.

Dedication

Στους γονείς μου, Μαίρη και Βασίλη

Preface

Foremost among all my teachers, Klaus Berkling taught me to question every assumption and not to rest until all aspects of a problem have been considered. For the last five years he has guided me through the maze of the practical aspects of Lambda Calculus reduction. With his judgment he reinforced my conviction that minimality and elegance are not luxuries but bare necessities.

Improving the practicality of Michael Hilton's *THOR/HORSE* was a strong influence and a well-defined first goal. It turned out that the problems are of a greater scope than I anticipated. Our interaction during the early stages of my study is greatly appreciated.

Inspiration, and the incentive for edging a bit closer to the "science" of computer science, was provided by the classes, seminars and talks that I was exposed to during my long career as a graduate student at Syracuse University. Ken Bowen, Lockwood Morris, John Reynolds, Alan Robinson, Ernest Sibert and Ed Storm come to mind, among many others. Academic committee members, Shiu-Kai Chin and Peter O'Hearn provided valuable feedback at every step.

Chuck Stormon and Coherent Research, Inc. afforded me the uncommon opportunity of continuing my graduate work while helping forge the future of this remarkable company. All my colleagues at CRi are to be thanked for their continuous interest, encouragement and good cheer.

Bruce Berra, Gideon Frieder and Brad Strait kindly provided financial assistance during my days at ECE, CIS and at the CASE Center. Their support is gratefully acknowledged.

From one-third across the globe away, loving parents Mairi and Vasilis were a vital source of encouragement, support and patience; enterprising sister Maria ensured that I stayed focused and balanced most of the time; Liebling Susanne E.(mail) Bohl was instrumental in making the journey more enjoyable. Lightheartedly, some credit also goes to a certain telecommunications company whose tariffs made it a bit more affordable to live and work in one continent while having so many of one's "friends & family" in another.

Table of Contents

1 LAMBDA CALCULUS AND REDUCTION COMPUTING 1

2 SYNTAX, NOTATION AND REPRESENTATIONS 25

List of Figures

List of Tables

Chapter 1

Lambda Calculus and Reduction Computing

1.1 INTRODUCTION - MOTIVATION

An intriguing direction in symbolic computing is to base an architecture (hardware or software) on equivalence preserving transformations of expressions. In this dissertation we show that this viewpoint can be made practical. The computational model is the Lambda Calculus—a simple model which views functions as rules. Our motivation is the belief that direct Lambda Calculus reduction can be done efficiently. Beta reduction in the scope of relative free variables is not difficult with proper representations. Strong normalization[1] opens up new applications by not relegating abstract interpretation to a meta level.

Berkling's Head Order Reduction (HOR) is the main tool of this approach. It is a delayed substitution algorithm with emphasis on the extended left spine. It uses the *de Bruijn* coding for variables, *unbound variable counts* for "neutralizing" relatively free variables and proceeds to an operational variant of the natural midway point of reduction, the *head normal form*. The machinery of Head Order Reduction appears to be a direct and minimal match to the requirements of strong reduction. Among its operational properties are extremely low space requirements and independence from pointer representations for variables. Michael Hilton's *THOR/HORSE,* an applied derivative system, is a good example of the overall direction and goals.

---

[1] In this thesis this term simply means that reduction proceeds to full β-normal form.

THOR/HORSE has demonstrated that systems based on the preceding principles are not only feasible but can also attain levels of efficiency previously thought unreachable. Hence, we downplay transformation techniques like combinators and strong typing that are commonly employed solely for the purpose of compilation to von Neumann code. It is our belief that specialized hardware or a low-level software "interpreter" for a proper operational casting of the Lambda Calculus can serve large classes of problems well, both in terms of expressiveness and efficiency.

But a pure normal-order strategy like that of *THOR/HORSE* is unacceptable for practical work if it is not augmented with sharing. A large part of our efforts concentrates on augmenting HOR with sharing. Initially, the task seemed intractable; early reducers were prone to subtle binding errors. As our understanding improved, we realized that no single technique is a complete solution to the problem and free of disadvantages. But progress has been made. By utilizing combinations of essential techniques provided by Berkling's earlier work the state of affairs described in this dissertation was reached. The initial goal of adding effective full sharing in a system having the scope of *THOR/HORSE* is unfulfilled.

The essentials of Head Order Reduction were implicitly present as early as 1972 in de Bruijn's *Indagationes Mathematicae* publication of a Lambda Calculus with integer "Nameless Dummies" instead of the usual named variable identifiers. Retrospectively, one can state that the logical progression that leads to the operational HOR is the addition of delayed substitutions via an environment and the emphasis on head normal forms. Historically, the development took a different course via Berkling's "in-the-large" reduction rules presented for the first time at the *Santa Fe Graph Reduction Workshop* in 1986.

The quest for grasping the interrelationships and intricacies of reduction methods led us to review several of the well-known weak and strong algorithms. Our attention was limited to the pure, untyped Lambda Calculus to keep complications to a minimum. It turned out that the weak form of HOR is identical in strategy with the so called "RTLF" strategy of Wadsworth and Aiello–Prini and other less well known algorithms. Detailed comparisons of all the popular algorithms demonstrate the directness and efficiency of HOR.

Our work represents a shift in the point of view of traditional research on functional programming. It is based on the realization that the Lambda Calculus, which is commonly regarded as one of the foundations of programming, is not intrinsically oriented towards producing a value. Therefore, not all languages and systems should be limited in this regard. We believe that much may be gained in the realm of software technology if the first class objects are full expressions rather than ground values of some output domain.

We attempt to derive architectural principles while remaining close to the inherent mechanisms suggested by the underlying theory. The sort of computations that benefit from the existence of a reduction system are those in which one is not only interested in getting a final ground result but also insists that the steps taken in arriving at such result are as transparent and direct as possible. If this line of thinking is followed, then the Lambda Calculus appears indeed to be a solid choice because most of the thorny issues surrounding declarative computation of the functional kind surface here in their purest form. How practical or influential this path may be is something that will be determined eventually but for the moment it seems exciting work to pursue.

Combinator techniques are sidestepped because we regard them as indirect. With combinators one forgoes the random reference ability of variables and replaces it by argument propagation systems. From an operational standpoint this is cumbersome and circuitous. In addition, such techniques make it very difficult to do *incremental* reduction and to produce recognizable intermediate results. The efficiency of reducers like HOR may be largely attributed to using binding indices for variables together with a correction mechanism for dynamically-created abstractions that operates in an incremental and local manner. This way one can avoid the problems of scope and name capturing of variables and thus render the detour via combinators much less attractive.

Our overall emphasis is, by necessity, the invention of reduction schemes which are economical from a practical rather than a theoretical standpoint. Hence, we reject the usual notion of optimality—which equates excess work to excess reductions—as too limiting and not very helpful for reduction system design. The Lambda Calculus can model situations of arbitrary complexity. A reduction strategy which always avoids excess reductions requires bookkeeping which is more substantial than the effort required for performing "excess" reductions. A more reasonable goal is to attempt to match or exceed the natural savings of applicative order while avoiding, if possible, its incompleteness. Opportunities for parallelism are plentiful, but again, the orchestration overhead can outweigh the benefits derived over simpler sequential tactics. Source program transformations can help in exposing the inherent parallelism and research in this area seems to us worthwhile.

The rest of this chapter highlights the reduction-based viewpoint and pinpoints the challenges that are present. Results are summarized in section 1.7.

1.2 OPERATIONAL LAMBDA CALCULUS AS COMPUTATIONAL/MACHINE MODEL

Attempting to derive architectural principles from the operational characteristics of Lambda Calculus reduction is a slightly unconventional line of research. It could be likened to an aspiration to advance the techniques of modern microprocessor design by examining closely the workings of a Turing Machine. But the parallel breaks down—and hence the former does not seem as unconvincing as the latter—if one considers that a Lambda Calculus expression is a higher level of expression than, let us improvise, the transitions rules of a

Turing Machine and that it should impart more structure to a mechanization procedure. Additionally, the declarative nature of the Lambda Calculus is in sharp contrast with the sequential instructions and state-oriented nature of the von Neumann model. Consequently, since the semantic gap to be covered is larger, the journey could be more rewarding.

Lambda Calculus as Basis of a Reduction System

Our interest in the Lambda Calculus stems from two basic premises: First, research efforts on making high-order declarative languages work better and serve larger classes of practical computing tasks should intensify[2]. Second, a significant portion of our current difficulties with the building of software is attributed at to the low-level nature of the popular programming language and environments and the batch style of most of real world development tools. Higher-order programming, symbolic computation, partial evaluation and incremental development hold serious promises in advancing common computing practice[3]. Programming environments could become more fluid and the so called "Computer-Aided Software Engineering" tools would be able to focus more on the actual artifact.

One could regard computation as proceeding under two rather distinct flavors. First, functions are applied to arguments in a pure sense by fusing them together which means propagating arguments to the proper places at the appropriate time. Second, ground data objects, like encodings for numbers, characters or graphics primitives, are transformed in well-defined and predesigned ways; for example, multiplying two integers together or shifting the value of a register by a fixed amount. Symbolic computation often has a stronger first component while "number crunching" is nearly a pure form of the second. The Lambda Calculus is one of the simplest settings where one can experience these two flavors in a distilled manner and offers the opportunity of discovering techniques that may be useful in more general settings. The considered application areas are reduction languages, semantics-directed machine architecture and high-level software development tools.

One cannot avoid touching upon the potential criticism that the Lambda Calculus is inherently a model of *sequential* computing and hence efforts based on it may not be as applicable to future systems[4]. Exploring concurrent computation in a similar spirit, starting

---

[2] We are particularly interested in function-based languages. But efficient Lambda Calculus reduction can have implications for languages with side effects due to the generality of the concepts of abstraction and application.

[3] The literature has a wealth of research on each of these topics and it is outside of the scope of this thesis to attempt further explications. For high-order functional programming see, for instance, the paper by J. Hughes [Hughes 90] and the proposal by J. Backus et al [Backus 90].

[4] The major difficulty is that from a foundational viewpoint the pure calculus can not "escape" an undefined argument of a disjunction; i.e., there is no mechanism for expressing

maybe with Milner's *Calculus of Communicating Systems* (*CCS*) [Milner 80] or Hoare's *Communicating Sequential Processes* (*CSP*) [Hoare 85] or some other concurrent process modeling theory, would have been closer to recent trends. The author feels though, that progress remains to be made at the level of sequential expression. As it will become evident in section 8.4, the core of the Head-Order strategy has important implications for automatic *run-time* parallelism given the structural invariance of the operational head normal forms.

An Example of Reduction Semantics

A reduction system is envisioned as a delivery platform for exploratory programming environments of declarative or well-structured imperative nature. Solid representation choices and direct support from a specialized instruction set can result in a high-level organization, compared to the von Neumann one. Compilation is limited to optimizing but equivalence-preserving transformations on portions of a program that consist of arithmetic, logical and other low-level hardware-supported operations.

To demonstrate this sort of system let us visit a well-known high-order function like *reduce*. In the syntax of *THOR, reduce* has a definition identical to that of a *Scheme* function. It also behaves identically to a *Scheme* function as seen in the first and second examples of Table 1.1. But a vital, and commonly considered prohibitively expensive, change of semantics is evident in the third and fourth runs. When *reduce* is supplied with fewer arguments than those implied by its definition or with enough but unbound ones, *HORSE* (the programming environment for *THOR*) is perfectly content to perform the available reductions without any of the usual error messages. This is akin to abstract interpretation but it is accomplished with nothing more than the ordinary reduction mechanisms of the underlying calculus; there is nothing "abstract" or "meta" about such behavior—it is simply that the rules of the underlying calculus are followed to a greater extent.

A major theme of our work is that the cost of such fluidity is not as high as commonly thought; the proviso is, of course, that efficient reduction algorithms and systems are available. Fundamentally, not much is new or radical with this approach. The novelty is the recognition of the importance of *expression-based* computation which advances beyond weak normalization.

---

the meaning of a "parallel or". Applied variants of the pure calculus like *THOR* can and do include such mechanisms by treating, for instance, truth values specially. At this level there is no such constraint; in fact, the side-effect free nature of the calculus makes it relatively easy to add parallelism. Graph transformations, the natural mechanism for reduction, do not impose any sequentiality by themselves.

THOR Definition	HORSE Run

`(define reduce`
`   (lambda (f id list)`
`      (if (null? list)`
`         id`
`         (f (car list)`
`            (reduce f id (cdr list)))))`
 	
`(define MS`
`   (lambda (r l f i) r f i l))`
	`horse> reduce * 1 [4 3 5]`
	`60`
	`91 reductions, total.`
	`Elapsed time: 0.002 seconds.`
	
	`horse>`
	`reduce (lambda (el len)(1+ len)) 0 [3 5 4]`
	`3`
	`67 reductions, total.`
	`Elapsed time: 0.002 seconds.`
	
	`horse> reduce f id [4 3 U 5]`
	`(F 4 (F 3 (F U (F 5 ID))))`
	`135 reductions, total.`
	`Elapsed time: 0.003 seconds.`
	
	`horse> MS reduce [4 3 u 5] f`
	`(LAMBDA (I) (F 4 (F 3 (F U (F 5 I)))))`
	`139 reductions, total.`
	`Elapsed time: 0.003 seconds.`

Table 1.1 In a reduction system arguments may be ground, unbound or even absent.

Sharing is Sine Qua Non

The total picture though is not as inviting as one might lightly think. Reduction semantics leads to concise, natural definitions which, under pure normal order evaluation, can result in utterly expensive computations if no mechanism for sharing intermediate results is present. Table 1.2 shows a definition of the Sieve of Eratosthenes in the spirit of Turner [Turner 79]. This definition utilizes infinite structures and a declarative functional style.

The *HORSE* run reveals what was just hinted. The situation is quite hopeless from a practical point of view; for each new computed prime number, normal-order evaluation leads to a repetition of all previous computation. Hilton provides in his thesis a similarly disappointing example based on a game tree. Numerous expressions will be shown later whose reduction costs vary widely depending on whether or not sharing is present.

This is a demonstration of the well known effect of "call-by-name". The evaluator itself adds one or more orders of complexity to the natural demands of the algorithm. In order to avoid this potential explosion in complexity, one needs to tightly orchestrate the cooperation of generator procedures like *from* with consumer procedure like *filter* and *sieve*. This is the crux of what is generally referred to as *laziness*. But it is easy to see that a judicious mechanism that always reuses intermediate results is hard to achieve in general.

THOR Definition	HORSE Run
| ```
(define primes
  (sieve (from 2)))

(define sieve
  (lambda (l)
    (let ((a (car l)))
      (cons a (sieve (filter a l))))))

(define filter
  (lambda (p l)
    (let ((a (car l))
          (b (cdr l)))
      (if (= 0 (modulo a p))
        (filter p b)
        (cons a (filter p b))))))

(define from
  (lambda (n)
    (cons n (from (1+ n)))))

(define take
  (lambda (n l)
    (if (= 0 n) NIL
      (cons (car l)
            (take (1- n) (cdr l))))))
``` | ```
horse> take 4 primes
[2 3 5 7]
1629 reductions, total.
Elapsed time: 0.035 seconds.

instructions executed: 4429
max graph nodes: 163
  max env size: 380
 max stack size: 52
   max aux size: 3

horse> take 9 primes
[2 3 5 7 11 13 17 19 23]
1108181 reductions, total.
Elapsed time: 17.588 seconds.

instructions executed: 2854732
max graph nodes: 551
    max env size: 1778
  max stack size: 165
    max aux size: 8
``` |

Table 1.2 Sieve of Eratosthenes. Normal-order reduction without sharing is unacceptable in a practical system.

Our work differs from other, more conventional, approaches in that it tries to reach this ideal without abandoning the structures of the Lambda Calculus in favor of transformation techniques. Usually, strong typing, restriction of output to ground value results and compilation to straight-line code via lambda lifting are utilized to achieve high performance.

1.3 CHARACTERISTICS OF REDUCTION COMPUTING

Expression-based computing does incur costs which are not easy to justify when manipulation of state is the *only* desired outcome. However, even though computationally demanding, it is a welcome addition as a semantic-level tool that is progressively more affordable as the delivery technology advances. Compilation, when seen as the general concept of perusing program text so that competent state-manipulation code can be produced, is still applicable. But as software writers we could do without the ubiquitous presence of the batch *edit-compile-run* cycle, mostly an historic artifact with little relation to present needs and machine capabilities.

The symbolic nature of expression computing based on the Lambda Calculus is of general *applicability*. This is because it is present in all parts of a program and is uniformly relevant to all programs. The term *abstract interpretation* is often used for transformation of expressions before ground arguments are supplied to them. In a Lambda Calculus reduction system there is no distinction between "execution" and "transformation". The symbolic capabilities are an inseparable part of the basis of the system since they are provided at the lowest level.

However, such expression transformations are not always of general *utility* because they are the result of one rule, namely β-contraction, and of a singular tactic namely the generation of a (strong) normal form. They may not always be by themselves what the problem requires. For example, partial evaluation may need to proceed in ways other than normal-order normalization. Therefore, even while most programming situations may benefit from the high-level of the underlying platform, some will require transformations which are different from the innate ones.

A narrow example of these thoughts is the following situation. In a reduction system based on the Lambda Calculus it is easy to "freeze" evaluation by providing a quantum that limits the number of reductions. But in any situation of some complexity there is no simple rule that determines the number of reductions an expression needs to reach a form which is interesting from the user's viewpoint. Similarly, when one is exploring behavior of functions interactively, it is often difficult to locate the interesting points in the evaluation; e.g., those which begin a specific function or a new recursive call.

But these difficulties are rooted more in our increased expectations for automatically purposeful behavior rather than in an inherent limitation. Meta-level mechanisms for controlling reduction need to be invented. It should not be very difficult to provide facilities like those found in modern functional/logic systems. Reduction systems based on techniques described in this thesis do have a partial remedy for problems of this sort. As we will see in section 4.5 the mechanisms of HOR provide for interactive expression navigation and selective incremental reduction.

Example of Reduction Programming

In a system where all computation results from manipulations that have simple semantics, the transparency of an algorithm or a transformation is high. That means that we can afford much higher confidence in programming solutions. Again, because of the generality of the Lambda Calculus as a foundation, such systems have wide applicability. Even areas which are considered removed from symbolic computations, e.g. arithmetic and scientific computing can benefit.

THOR Definition	HORSE Run
| $$\sin(x) = \sum_{i=0}^{\infty} (-1)^{i} \frac{1}{(2i+1)!} x^{2i+1}$$<br><br>```
(define horners-rule
  (lambda (coefficient variable length)
    (letrec
      ((term
        (lambda (n)
          (if (= n length)
            (coefficient n)
            (+ (coefficient n)
              (* variable (term (1+ n)))))))))
    (term 0))))

(define sine-co
  (lambda (i)
    (/ (expt -1.0 i)
      (factorial (+ (* 2 i) 1)))))

(define sine-series
  (lambda (x x2)
    (* x (horners-rule
      sine-co x2
      (sine-number-of-terms epsilon
        (/ pi 2)))))))

(define sine-half
  (lambda (x)
    (let ((x2 (* x x)))
      (sine-series x x2))))
``` | ```
horse> horners-rule C V 10
(+ (C 0) (* V (+ (C 1) (* V
(+ (C 2) (* V (+ (C 3) (* V
(+ (C 4) (* V (+ (C 5) (* V
(+ (C 6) (* V (+ (C 7) (* V
(+ (C 8) (* V (+ (C 9) (* V
  (C 10)))))))))))))))))))))
58 reductions, total.
Elapsed time: 0.003 seconds.

horse> horners-rule sine-co V 5
(+ 1.000000
  (* V (+ -0.166667
    (* V (+ 0.008333
      (* V (+ -0.000198
        (* V (+ 0.000003
          (* V -0.000000)))))))))))
309 reductions, total.
Elapsed time: 0.010 seconds.

horse> sine-half
(LAMBDA (X)
  (* X (+  1.000000 (* (* X X)
    (+ -0.166667 (* (* X X)
    (+  0.008333 (* (* X X)
    (+ -0.000198 (* (* X X)
    (+  0.000003 (* (* X X)
      -0.000000)))))))))))))

760 reductions, total.
Elapsed time: 0.025 seconds.
``` |

Table 1.3 (Hilton) Example of the source-transformation capabilities of a reduction system.

An illustration of the previous view is attempted in Table 1.3. This example is provided by Hilton in the manual for *THOR/HORSE* [Hilton 90a]. It was adapted from a paper by Gerald Roylance [Roylance 88]. The core idea is that a numerical function like *sine* can be programmed constructively by transforming its mathematical definition rather than simply as a loop which specifies changes to state variables.

The left part of Table 1.3 shows four of the functions that describe symbolically the process of computing *sine* via the expansion of a specific Taylor series. In particular the first function defines *Horner's Rule* for efficient evaluation of a series. The second function returns the coefficient of each term in the overall sum The last two rules put everything together by defining how *sine* is to be calculated over the limited interval $[-\pi/2, \pi/2]$.

The right part of Table 1.3 shows interactive reduction of some portions of the definition of *sine*. The first one affords us confidence that Horner's rule is indeed programmed correctly. An invocation with the actual coefficients and with a specific number of terms returns an expression which is an approximation of the final solution. Finally, the last run not only produces an expression which is ready to give a value result—

when and if it is applied to a numeric argument—but it is also proof of a numerical approximation that suffices for computing *sine*. The cost of the "symbolic" derivation, however minimal, does not have to be incurred each time *sine* is used in a numerical context; the result for *sine-half* could be assigned—*rdefined* in *HORSE* parlance—to a new symbol for subsequent use.

To summarize, our central concern is to invent reduction algorithms that are simple enough to be generally applicable and include the highest affordable degree of sharing.

1.4 DIRECT LAMBDA CALCULUS REDUCTION

It was mentioned previously that we regard combinator reduction techniques as indirect and not as suitable as direct lambda reduction to be the basis of a reduction system. To visually reinforce the indirectness of Combinatory Logic (*CL*) reduction we present an example. Figure 1.1 shows normal order reduction sequences for

$$(\lambda a.(a\ \lambda b.(b\ a\ \lambda cd.d\ \lambda ef.f\ \lambda gh.(b\ g\ (g\ h)))\ \lambda ijkl.k)\ (\lambda uv.(u\ (u\ v)))),$$

an expression which computes the predecessor of the second *Church Numeral*[5]. Part (*A*) shows a combinator reduction starting with the above expression after it has been transformed to a combinator expression via a suitable abstraction algorithm[6]. Part (*B*) shows the corresponding direct lambda reduction in *BTF* notation (*BTF* is described in section 2.3). Part (*C*) presents, in a mixed calculus starting with the *CL nf*, the extra reductions needed to reach the final strong normal form. Part (*D*) lists in *BTF* notation the lambda definitions of the combinators used in this exercise. The *BTF* reduction sequence with the redices highlighted is shown also in Table B.1.

The pattern that emerges is typical of the behavior of combinators. Even though the "grain" of a single combinator reduction is large (typically, one to four arguments are absorbed simultaneously) substantially more reductions are required for doing the same amount of work[7]. This is a direct result of the process of abstraction; a large percentage of

---

[5] $C_n = \lambda xy.\underbrace{x(x(...(xy)...))}_{n\ times}$. See, [Barendregt 84], p140.

[6] The abstraction algorithm and combinator set employed are those presented by K.J. Greene in the development of the *LNF* combinator reduction system. For details, see his dissertation [Greene 85].

[7] Combinator reduction does have the benefit of effectively performing η-reduction as part of the abstraction process.

C.

```
(W (B B (K I)))
(-1 B B (K I) 0 0)
(-1 (-2 B (1 0)) (K I) 0 0)
(-1 (-1 B (K I 0) 0) 0)
(-1 B (K I 0))
(-1 (-2 K I 2 (1 0)) 0 0)
(-2 K I 1 (1 0))
(-2 (-1) I (1 0))
(-2 1 (1 0))
(-2 1 0)
```

D.

```
S  = (-3 2 0 (1 0))
C  = (-3 2 0 1)
CC = (-4 3 (2 0) 1)
R  = (-2 0 1)
W  = (-2 1 0)
B  = (-3 2 (1 0))
K  = (-2 1)
I  = (-1 0)
```

A.

```
(C (S I) (CC S (CC C (CC C R (K I)) (B W (B B))) (K (K K)) (W B))
(S I (CC S (CC C (CC C R (K I)) (B W (B B)) (W B) (K (K K)))
(I (W B) (CC S (CC C (CC C R (K I)) (B W (B B)) (W B) (K (K K)))
(W B (CC S (CC C (CC C R (K I)) (B W (B B)) (W B) (K (K K)))
(CC S (CC C (CC C R (K I)) (B W (B B)) (W B) (CC S (CC C (CC C R (K I)) (B W (B B)) (W B) (K (K K)))
(CC C (CC C R (K I)) (W B) (CC S (CC C (CC C R (K I)) (B W (B B)) (W B) (K (K K)))
(CC C (CC C R (K I)) (W B) (CC S (CC C (CC C R (K I)) (B W (B B)) (W B) (K (K K)))
(C (R (W B)) (K I) (CC S (CC C (CC C R (K I)) (B W (B B)) (W B) (K (K K)))
(R (W B)) (K I) (CC S (CC C (CC C R (K I)) (B W (B B)) (W B) (K (K K)))
(CC S (CC C (CC C R (K I)) (B W (B B)) (CC S (CC C (CC C R (K I)) (B W (B B)) (W B) (K (K K)))
(CC C (CC C R (K I)) (W B) (K) (K (K K)) (B W (B B)) (W B) (K (K K)))
(CC C R (K I)) (W B) (K (K K)) (B W (B B)) (W B) (K (K K)))
(C (R (W B)) (K I) (K (K K)) (B W (B B)) (W B) (K (K K)))
(R (W B)) (K I) (K (K K)) (B W (B B)) (W B) (K (K K)))
(K K) (W B) (B W (B B)) (K (K K)))
(K K) (K I) (B W (B B)) (K (K K)))
(I (W B)) (B W (B B)) (K (K K)))
(I (K I) (B W (B B)) (K (K K)))
(B W (B B) (B W (B B)) (K (K K)))
(W (B B (K K)) (K I) (B W (B B)) (K (K K)))
(W (B B (B W (B B)) (K (K K)))
(W (B B (R (W B)) (K I) (K (K K)))
(W (B B (K K) (K I) (B W (B B)) (K (K K)))
(W (B B (K K I) (K I) (B W (B B)) (K (K K)))
(W (B B (K I) (B W (B B) (K (K K)))
(W (B B (K I))
```

B.

```
((-1 0 (-1 0 1 (-2 0) (-2 2 1 (1 0)) (-4 1) (-2 1 (1 0)))
(-2 1 (1 0)) (-1 0 (-2 1 (1 0)) (-2 0) (-2 2 1 (1 0)) (-4 1))
(-1 (-1 0 (-2 1 (1 0)) (-2 2 1 (1 0)) (-2 0) (-2 2 1 (1 0)) (-4 1))
(-1 0 (-2 1 (1 0)) (-4 1) (-2 1 (1 0)) (-2 0) (-2 2 1 (1 0)) (-4 1))
(-1 0 (-2 1 (1 0)) (-4 1) (-2 0) (-2 0) (-2 2 1 (1 0)) (-4 1) 1 (1 0)))
(-4 1) (-2 1 (1 0)) (-2 0) (-2 0) (-2 4 1 0) (-2 1 (1 0)) (-4 1) 1 (1 0))
(-3 1) (-2 0) (-2 4 1 1 (1 0)) (-2 0) (-2 -1 0 -2 1 (1 0)) (-4 1) 1 (1 0))
(-2 1) (-2 0) (-2 -4 1) 1 (1 0) (-2 0) (-2 2 1 (1 0)) (-4 1) 1 (1 0))
(-3 0) (-2 -4 1) 1 (1 0) (-2 0) (-2 -1 0 -2 1 (1 0)) (-4 1) 1 (1 0))
(-1 0) (-2 0) (-2 -1 0 -2 1 (1 0)) (-2 0) (-2 2 1 (1 0)) (-4 1) 1 (1 0))
(-2 0) (-2 0) (-2 -1 0 -2 1 (1 0)) (-2 0) (-2 2 1 (1 0)) (-4 1) 1 (1 0))
(-2 0) (-2 0) (-2 0) (-2 2 1 (1 0)) (-4 1) 1 (1 0))
(-2 -1 0 -2 1 (1 0)) (-2 0) (-2 2 1 (1 0)) (-4 1) 1 (1 0))
(-2 -3 1) (-2 0) (-2 -4 1) 1 (1 0) 1 (1 0))
(-2 -2 1) (-2 0) (-2 -4 1) 1 (1 0) 1 (1 0))
(-2 -3 0) (-2 -4 1) 1 (1 0) 1 (1 0))
(-2 -2 0) 1 (1 0)
(-2 -1 0) (1 0)
(-2 1 0)
```

Figure 1.1 Combinator reduction contrasted to direct lambda reduction.

the reductions accomplishes nothing more than the propagation of arguments to their problem-originated context—i.e., simply the role played naturally in the Lambda Calculus by the random reference ability of variables. Admittedly, the runtime mechanisms for an implementation of *CL* reduction are simpler since the difficulties associated with variables are absent. But, as we shall see, these difficulties are not as prohibitive as they are commonly thought to be.

The abstraction process itself is not without costs. In fact, for an expression *E* of significant size in relation to the number of reductions required for *E* to reach *nf,* the cost of abstraction can easily exceed the cost of direct reduction. This point is not elaborated any further but it should not be surprising since both abstraction and normalization require, at minimum, a complete traversal of the expression at hand. In general, it is known that abstraction results in the worst case to quadratic expansion and in the average, via suitable encodings, to expressions of size $O(n \log k)$ where n is a measure of the size of the input and k a measure of the total number of abstractions present [Kennaway 88].

Admittedly, the potential benefits of sharing that are obtained via *graph* combinator reduction are not evident in the last example of fully-effected reductions and a linear representation. Our tenet is that the potential benefits of sharing are of similar degree and equally easy to apply to both Calculi if suitable machinery is invented.

Our last comment is about transparency. Reduction in *BTF* is not exactly readable but it enjoys a trivial correspondence to lambda reduction with named variables. A combinator abstraction process on the other hand does not have any transparency because a vital aspect of the original expression is absent. It can hardly be claimed that any similarity to the original problem expression can be maintained without substantial annotating efforts.

In a reduction system like *HORSE,* variable names are propagated throughout during the reduction process at little additional cost. Transparency is maintained at all times, even during incremental reduction. A mechanism of *protecting* named variable occurrences from the immediately enclosing binder is employed. This technique is rooted to proposals by Berkling [Berkling 82]. It consists of the addition of an unbinding operator which, when placed in front of a variable occurrence, "neutralizes" the effect of an enclosing abstractor having the same bound variable as that of the protected variable. It is based on the convention that *x* occurs free in *λx.#x* but bound to the outer abstractor in *λx.(λx.#x)*. The same mechanism is also employed in the *π-RED* series of systems.

1.5 APPLIED REDUCTION SYSTEMS

True reduction systems are a relatively rare genre at this moment in time. The ones that we are aware of have been influenced by Berkling's research. As already mentioned,

Hilton's *THOR/HORSE* is based on Head Order Reduction and implements a pure normal-order strategy [Hilton 90a], [Hilton 90b]. The *π-RED* series of systems from the University of Kiel [Schmittgen 92] are based on graph reduction of an applied extension of the Lambda Calculus and implement an applicative-order policy. Evaluation order notwithstanding the two systems appear to achieve similar performance levels judging from the few common tests that we attempted (cf. section 7.6).

Both *HORSE* and *π-RED* offer incremental reduction, i.e., the ability to specify the maximum number of reductions that are to be performed. In addition, the *π-RED* systems feature a full-screen, syntax-directed editor which allows any subexpression to be targeted for reduction. This ability is also naturally provided by the HOR techniques; see section 4.5 and the system described in Appendix C.

Table 1.4 shows the definition and reduction of one of our popular test expressions in the *π-RED* system. The term $(\lambda a.(a\ \lambda b.(b\ a\ \lambda cd.d\ \lambda ef.f\ \lambda gh.(b\ g\ (g\ h)))\ \lambda ijkl.k)$, which computes

```
┌─────────────────────────────────────────────────────────────────────────────────┐
│ OREL Definition & π-RED* Run of the Church Numeral Predecessor Function           │
╞═══════════════════════════════════════════════════════════════════════════════════╡
│ Input   :                                                                          │
│ ---------+---------+---------+---------+---------+---------+---------+---------+------│
│ ap sub [ A ]                                                                       │
│    in ap ap A                                                                      │
│          to [ sub [ B ]                                                            │
│                in ap ap ap ap B                                                    │
│                            to [ A ]                                                │
│                            to [ sub [ C , D ]                                      │
│                                   in D ]                                           │
│                          to [ sub [ E , F ]                                        │
│                                 in F ]                                             │
│                        to [ sub [ G , H ]                                          │
│                              in ap ap B                                            │
│                                     to [ G ]                                       │
│                                    to [ ap G                                       │
│                                        to [ H ] ] ] ]                              │
│        to [ sub [ I , J , K , L ]                                                  │
│               in K ]                                                               │
│ to [ sub [ U , V ]                                                                 │
│      in ap U                                                                       │
│         to [ ap U                                                                  │
│             to [ ap U                                                              │
│                 to [ ap U                                                          │
│                     to [ ap U                                                      │
│                         to [ ap U                                                  │
│                             to [ V ] ] ] ] ] ] ]                                   │
│                                                                                    │
│ Output  :                                                                          │
│ ---------+---------+---------+---------+---------+---------+---------+---------+------│
│ sub [ G , H ]                                                                      │
│ in G [ G [ G [ G [ G [ G [ H ] ] ] ] ] ]                                           │
│                                                                                    │
│ Message : Performed 22964 reduction steps, Processing time : 1.62 seconds          │
└─────────────────────────────────────────────────────────────────────────────────┘
```

Table 1.4 University of Kiel π-RED system on the pred7 test.

the predecessor of a *Church Numeral* strictly in the pure Lambda Calculus, is shown again. This operator is applied to the lambda term $\lambda uv.u\,(u\,(u\,(u\,(u\,(u\,(u\,v))))))$, representing the number seven, and the result is the term $\lambda uv.u\,(u\,(u\,(u\,(u\,(u\,v)))))$, a coding of the number six.

The ability to express an operation in the base calculus affords us confidence that in case it is decided to include it as a primitive in a practical system, purely for engineering reasons, no semantic principles are violated. A simple but practically relevant example is the following.

One can define as a primitive an operator that acts like a lambda *selector* or a generalized *K* combinator. An expression which reduces to such a selector and which relies only on an equality test on integers in addition to basic arithmetic operators is shown in Table 1.5. This table lists a *THOR* definition and gives a trial *HORSE* run.

Incidentally, the second input expression of this run also demonstrates how the *C* combinator together with strong reduction can effectively swap the arguments before they

THOR Definition	HORSE Run
$S_k^n = \lambda nk.Y\;\lambda Bi.$ $\quad if\;(i = k+1)$ $\quad\quad \lambda z.Y\;\lambda Ai.$ $\quad\quad\quad if\;(i = 0)$ $\quad\quad\quad\quad z$ $\quad\quad\quad\quad \lambda x_a.\,A\,(i-1)$ $\quad\quad\quad (i-1)$ $\quad\quad \lambda x_b.\,B\,(i-1)\;\;n$ ```(define sel
  (lambda (n k)
    (rec
      (lambda (mklb i)
        (if (= i (+ k 1))
          (lambda (z)
            ((rec
              (lambda (mkla i)
                (if (= i 0)
                  z
                  (lambda (xa)
                    (mkla (- i 1))))))
            (- i 1)))
          ((lambda (xb) (mklb (- i 1)))))))
    n))``` | ```horse> (sel 9 5)
(LAMBDA (XB XB XB Y XA XA XA XA XA) Y)
73 reductions, total.
Elapsed time: 0.003 seconds.

horse> (C sel 5 9)
(LAMBDA (XB XB XB Y XA XA XA XA XA) Y)
77 reductions, total.
Elapsed time: 0.003 seconds.

horse> C
(LAMBDA (X Y Z) (X Z Y))
1 reductions, total.
Elapsed time: 0.000 seconds.

horse> (sel 5 3 a4 a3 a2 a1 a0)
A3
45 reductions, total.
Elapsed time: 0.002 seconds.

horse> (rdefine sel5-3 sel 5 3)
(LAMBDA (XB Y XA XA XA) Y)
40 reductions, total.
Elapsed time: 0.002 seconds.
SEL5-3 defined.

horse> (sel5-3 a4 a3 a2 a1 a0)
A3
6 reductions, total.
Elapsed time: 0.000 seconds.``` |

Table 1.5 A generalized selection combinator expressed as a lambda expression in a calculus with an equality test for integers.

become available to a function. This last point is often a source of confusion about Lambda Calculus based systems. A common assumption is that since, at the level of the calculus, arguments are supplied and consumed in the textual order of appearance, the user would face similar constraints at the level of programming.

For the curious reader we iterate that the equality test on Church Numerals is itself definable in the pure calculus. See Table 7.11 for a *THOR* definition and an example run.

1.6 APPLICATIONS OF REDUCTION COMPUTING

Since the Lambda Calculus is so centrally placed with respect to computing direct and efficient reduction has many applications. A reduction system is by its nature a partial evaluator. Strong normalization among other features allows data structures to take a "functional" character; they can be coded as head normal forms and combined or taken apart by applying them to lambda-defined operators. Such constructions can be used to both define the meaning and at the same time employed as efficient realizations (cf. section 8.3). Reduction can give a new dimension to "Software Engineering". It can help with the progressive refinement of specifications by making their formalization executable. Finally, reduction systems can be excellent pedagogical tools.

Below we list some other application areas. In the following subsection we sketch a possible use of a reduction system in exploring models of parallel processing.

- Direct reduction as basis of a high-level computing architecture

- Symbolic execution for development and testing

- Programming language research including object orientation

- Constructive development of algorithms

- Program verification and theorem proving

- Semantics-directed optimizations

- Parallel processing and data-parallel functional programming

- Explicit substitutions is an alternative for combinator graph reduction

Process Formalization

The availability of an efficient reduction system can be of consequence when experimenting with the expressions of a process algebra. As an example, the meaning of most of the rules of the Communicating Sequential Processes (*CSP*) theory has been presented in a pure lambda notation [Hoare 85]. With strong normalization and the intermediate operational techniques presented in the sequel, both abstract expression combinations and specific instances with concrete arguments can be manipulated in an equivalence-preserving manner revealing the essence of the required processing and exposing parallelism at the level of semantics.

To give the flavor of such usage, the defining lambda expressions for three of the basic operators of *CSP* are listed below. If the entire theory is captured by a set of similar definitions, one should be able to symbolically compose processes, selectively reduce the resulting expressions and run them on process definitions with concrete ground events.

$$prefix \ c \ P \ \equiv \ \lambda x. \ if \ (x = c) \ then \ P \ else \ \downarrow$$
$$choice \ c \ P \ d \ Q \ \equiv \ \lambda x. \ if \ (x = c) \ then \ P \ else \ (if \ (x = d) \ then \ Q \ else \ \downarrow)$$
$$concurrent \ P \ A \ Q \ B \ \equiv \ Y \ \lambda C. \lambda x. \ if \ (and \ x \in A \ x \in B) \ then \ (C \ Px \ Qx) \ else$$
$$(if \ (x \in A) \ then \ (C \ Px \ Q) \ else$$
$$(if \ (x \in B) \ then \ (C \ P \ Qx) \ else \ \downarrow)))$$

$$where \quad c, d: events ,$$
$$P, Q: processes ,$$
$$A, B: process \ alphabets.$$

It is the essence of the *CSP* theory that process expressions are higher order objects which, given an event as argument, produce as result a process expression which is type-wise indistinguishable from the original process. In other words, processes are infinite objects whose behavior is explored finitely via reduction and thus are best modeled through either an untyped calculus or one that allows infinite types.

1.7 RESULTS - TOUR

Highlights and Results

In this thesis the insights and improvements on existing reduction techniques are numerous but often subtle; notation and representations are nearly always the vital enabling elements. Nonetheless, it is customary to attempt an enumeration.

The emphasis on reduction as a powerful, general-purpose computing paradigm as well as the technical advances of Head Order Reduction is a life-long effort of Klaus Berkling. Michael Hilton's thesis [Hilton 90b] provides a tractable semi-formal account of the basis of Head Order Reduction and a solid treatment of an applied system that puts it in use. Within the above context, the following items are worth highlighting as specific contributions of this author. I have:

- Coded most of the influential Lambda Calculus reduction algorithms using a uniform representation based on simple tagging and either de Bruijn indices or Wadsworth pointers for variables. Attempted a taxonomy of the organizational themes and optimization techniques which are employed. Exercised them on various test expression sets in order to gain insight in their behavior and relative performance. Representative reducers for many of the above families described in mostly pure *Lisp* are found in Appendix A.

- Introduced a novel linear notation for casting lambda expressions, and run-time elements like closures/suspensions, unbound variable counts, etc. It is a natural extension of the "integer form with parentheses" employed by Berkling in his research for many years [Berkling 87]. Using this notation the weak, head and strong normal forms (*whnf, hnf* and *nf*) are given compact, operationally-oriented descriptions. In addition, this notation has close correspondence to independently introduced systems that formalize the notion of substitution.

- Described and analyzed the mechanisms by which correct and efficient sharing is accomplished in Berkling's HOR family. The resulting near normal-order reducer is one with perhaps the best overall (space/time) behavior and reduction count to date. This dramatic improvement over the classic non-sharing HOR is accomplished through the simple but vital addition of Berkling's mechanism for closing expressions artificially via the introduction of context tagging which signifies a suitable "dummy" environment. Subexpressions referred to by a head variable are reduced in isolation and the environment or redex store is selectively updated as described by Burge and others (cf. with the *Assigning Dump* for the delaying version of the *SECD* machine [Burge 75]). But our contribution is such that the latter is accomplished in general contexts that include relatively free variables without any preprocessing being necessary. The resulting call-by-need reducer improves on applicative-order reduction in terms of economy. However, because reduction of some subterms is attempted out of context, it does not manage to entirely avoid incompleteness.

- Described a sharing version of a weakly normalizing HOR reducer which compares very favorably with applicative-order *whnf* reducers. It is identical in overall strategy (but simplified operationally) to Wadsworth's *RTLF* strategy as captured by Aiello–Prini's interpreter [Wadsworth 71], [Aiello 81]. The convergence of these algorithms is remarkable and interesting on its own right.

- Succeeded in combining the *RTNF/RTLF* (exposition of the top abstraction) strategy of Wadsworth and the delayed substitutions and environment organization of Aiello and

Prini (*AP*) with HOR-style elements. This resulted in decreases of memory utilization for this family of reducers. The family, named *AP-HOR*, enjoys a very concise definition. Sharing in the example of Aiello and Prini can also be achieved in this version. Also, it is shown that the *HOR*-inspired optimization of "early variable-argument lookup" is applicable to this and the original pointer-variable *AP* reducer. The *RTNF/RTLF* strategy is generally a sound choice but it is more complicated operationally than HOR reduction. The sharing achieved with weak reduction of operator expressions is not as tight as the one achieved when proceeding from a head normal form; therefore, these reducers tend to perform more reductions than the sharing versions of *HOR*.

- Showed how HOR relates to de Bruijn's exposition of binding indices [Bruijn 72], and two converging theories which can be viewed partially as formalizations of the HOR machinery developed earlier by Berkling. The theories are Field's ΛCCL [Field 91], and Abadi-Cardelli-Curien-Lévy's "Explicit Substitutions", also known as the $\lambda\sigma$ -calculus [Abadi 91]. Together with our work they provide an indication that the quest for improved reduction engines for the Lambda Calculus has reached a new plateau.

- Made space/time/control complexity comparisons between distilled abstract versions of the above mentioned *whnf*, *hnf*, and *nf* reduction algorithms with emphasis on the relative performance of numerous variations of HOR. Some conclusions were reached as a result of this, inherently limited, testing. Also performed few space/time comparisons between three implementations of expression-based reduction systems under Unix (π-*RED** and π-*RED*+ from the University of Kiel [Kluge 93] and the *HORSE* system developed by Michael Hilton, formerly at Syracuse University [Hilton 90a]).

- Produced with Berkling an extensive experimentation environment for reduction in the Lambda Calculus. This system is called BTRD and is a stand-alone Macintosh application.

Historic Lineage

Over the last four decades there has been a lot of interest in the Lambda Calculus as a mathematical theory, as a tool for assigning meaning and, occasionally, as a model for machine computation. Even though the literature is voluminous only few works have surfaced which deal with the practical aspects of reduction *per se* with adequate operational detail. A common theme in the practically-oriented computer science community is to stray from some of the more difficult and wholesome aspects of the theory in the interest of efficiency. With our work we attempt to show that some of these deviations are not as necessary.

Although a bit risky, a timeline of contributions is presented in Figure 1.2. It starts with the origin of the calculus and proceeds to recent years. Such a listing is by necessity incomplete and subjective, focusing as it is on theories, texts and systems which are closest to our approach.

Combinatory Logic	24	Schöfinkel
	30	Curry
Lambda Calculus	41	Church
Combin. Logic Treatise	58	Curry-Feys
Eval-Apply, Recursive Programming	60	McCarthy, Dijkstra
SECD Machine , ISWIM	64	Landin
Normal/Lambda Graph Reduction	71	Wadsworth
"Nameless Dummies"	72	de Bruijn
Reduction Architectures	74	Berkling
Lazy SECD Machine	75	Burge
Hyper Pure Lisp	76	Henderson-Morris
FP	78	Backus
SASL, Combinator Reduction	79	Turner
Parallel Reduction, Optimality	80	Lév
Env. based RTNF/RTLF, LC Treatise	81	Aiello-Prini, Barendregt
Categorical Abstract Machine	85	Cousineau-Curien-Mauny
Spine Strategies	86	Barendregt et al
Head Order Reduction	87	Berkling
πRED Reduction System	89	Kluge et al
λσ-calculus, ΛCCL, THOR-HORSE	90	Abadi et al, Field, Hilton

Figure 1.2 Contributions to the theory and the practice of Lambda Calculus Reduction.

Class	Eval'd	Researcher(s) - Year	Description	Short Name
whnf	•	McCarthy 1960	Pure Lisp, eval-apply.	*EV-AP*
	•	Landin 1964	*SECD* machine.	*SECD-V*
	•	Landin 1964, 1966	*SECD*, normal order strategy.	*SECD-N*
		Burge 1975, Henderson–Morris, Friedman–Wise 1976	*SECD* normal and lazy strategy (Assigning Dump). Also, lazy applicative-order systems.	*SECD-L, HyperLisp, lazy cons*
	•	✳ , Oberhauser 1986	Weak Head Order Reduction.	*W-HOR*
	•	✳ , see also Aiello–Prini 1981	As above with sharing, identical in strategy to Aiello–Prini's RTLF.	*W-HOR-SH RTLF-SH*
		Cousineau–Curien–Mauny 1985	Categorical Abstract Machine.	*CAM*
hnf	•	✳	Operational Head Normal-HOR.	*HN-HOR*
nf		Wadsworth 1971	Pointer variables, full substitution, *RTNF/RTLF* strategy.	*WADS-N*
	•	Hilton–Berkling 1992	Non-deterministic abstract HOR.	*HOR-NS*
	•	Berkling 1986, Hilton 1990	Head Order Reduction; dB indices, stack, environment and *UVC*s.	*HOR, HORSE*
	•	✳	As above with sharing. Variations.	*HOR-SH*
	•	Aiello–Prini 1981	Delayed *RTNF/RTLF* strategy.	*AP*
	•	Aiello–Prini 1981	As above with sharing.	*AP-SH*
	•	✳	Breadth-First HOR.	*BF-HOR*
	•	✳	AP strategy HOR style, a force flag, dB indices and *UVC*s.	*AP-HOR*
	•	✳	As above with sharing.	*AP-HOR-SH*
	•	Schmittgen–Blödorn–Kluge 1992, et al	Applicative-order expression reduction, fully incremental system.	*π-RED*
NOTE	✳	: developed by the author and K.J. Berkling and first presented in this thesis.		

Table 1.6 Classification of reduction algorithms/systems and historical summary.

Reducer Classification

Next, a taxonomy of popular reduction algorithms is attempted. Table 1.6 groups influential algorithms by the normalization class (weak, head or strong) to which they naturally belong. It is also noted whether an algorithm has been scrutinized during the course of our study. We have placed among them our own contributions, appropriately classified. The reader will notice that some of the evaluation mechanisms listed in this table were not necessarily intended or first presented as reduction mechanisms for the Lambda Calculus. It is felt though that it is proper to include them in this table because they capture essential elements of reduction and of the issues that arise in implementations.

Chapter Tour

The organization of the rest of this thesis is as much a result of our quest to present a relatively self-contained and reasonably comprehensive account of Lambda Calculus reduction as it is an attempt to recount our research contributions.

Chapter 2, **"Syntax, Representations and Notations"**, reviews the basics of the Lambda Calculus, discusses the representations we have settled on for our abstract reducers and presents our notational device for direct reduction. Heavy use of this notation is made throughout.

Chapter 3, **"Graph Reduction Algorithms"**, presents influential graph reduction algorithms using the conventions of the second chapter. It starts with McCarthy's *Eval-Apply* and culminates with Berkling's Head Order Reduction (HOR). It is shown that HOR results from a number of simple but fundamental observations about what is needed in Lambda Calculus graph reduction. The relation between the in-the-large rules and the state transitions of the HOR abstract machine is explained.

In Chapter 4, **"Facets of Head Order Reduction"**, the focus is on the operational aspects of HOR. The important construction of a lazy head normal form is explained. We show how the HOR machinery is used to navigate expressions in an equivalence-preserving manner while performing reductions in subterms. After a sketch of the arguments that assure correctness, examples of how the HOR techniques can be used with alternative strategies are given. One of these strategies, a conservative "Breadth-First" HOR which still obeys normal order evaluation, may have significant implications for automatic parallelism.

In Chapter 5, **"Related Theories"**, the similarities of the operational casting of Head Order Reduction to de Bruijn's original presentation of binding indices, the theory of Explicit Substitutions and to J. Field's ΛCCL calculus are touched upon briefly.

Chapter 6, **"Sharing Strategies"**, focuses on the vital issue of adding effective sharing in HOR-class reducers. After reviewing the techniques that are employed in related reducers, a mechanism is presented by which correct sharing is achieved in HOR. We demonstrate the practical significance of this development and explain its theoretical weakness.

Chapter 7, "**Comparative Analysis and Evaluation**", reports on the results of comparisons of the main reduction strategies on common test expression sets focusing on comparisons of HOR to previous algorithms.

Finally, in Chapter 8, "**Towards Practical Lambda Calculus Systems**" we pick up the discussion of the goals set forth in the first chapter. The issues raised by low-level realizations of HOR, the introduction of arithmetic and other primitives, and the demands of sharing are visited briefly. The implications of Breadth-First HOR for parallel evaluation are exemplified. We conclude this dissertation with a discussion of directions for further research.

Appendices A and B provide concrete definitions for most of the discussed abstract reducers and show full traces of the more elaborate examples.

Appendices C and D consist of a short user's guide for our experimentation system and a listing of the main set of test expressions employed in our statistical evaluations.

Remarks

In most of the literature, efficiency of reduction is viewed solely in terms of its length. Minimum length sequences are defined as optimal. Our work so far has led us to believe that, pragmatically speaking, the question of optimality of reduction in the Lambda Calculus is not by itself a important one. The "unit value" and the "unit cost" of accomplishing one reduction are too small to justify elaborate mechanisms that may require complicated decision procedures or long-winded knowledge of a machine's state. If a reducer is to be considered practical it has to implement a relatively simple policy that works well under most but, understandably, not all circumstances. Once gross inefficiencies have been removed, optimality with respect to minimizing the number of reductions does not appear to correlate well with operational economy[8]. A universally optimal strategy would be tantamount to a global optimization procedure and hence there is no hope that one can be had at acceptable costs.

Formal results appear to support this viewpoint. Optimal strategies have been proved non-recursive when expressions without a normal form are not excluded. If we exclude such expressions, exhaustive enumeration of all reduction sequences with length equal or shorter to that of normal-order could identify the sequences with minimal length; naturally such a tactic is of no practical significance.

---

[8] A lower reduction count for achieving normal form is often more of a sign of "conservativeness" rather than of overall economy. Substantial decreases in reduction counts however usually correlate well with efficiency.

These views appears to be shared by other investigators—indeed even by some among those who have attempted to invent optimal algorithms. Reduction algorithms which purport to achieve optimality share a common characteristic which, given the remarks of the previous paragraph, appears to disqualify them from practical considerations:

> … I have shown, however, that ΛCCL is insufficient for implementing optimal reduction schemes and thus more than shared closures, environments, or λ-terms are apparently necessary if optimality is to be achieved at all. … [Field 90], p.13.

> … Theorem A2. For any $k > 0$, there is no recursive reduction strategy that can reduce any λ-expression to normal form (if it exists) in no more than $k \times m$ number of steps, where m is the minimum number of steps needed to reduce the expression. … [Kathail 90], p.190.

> … Schemes, by Staples [Staples 80] and, recently, by Lamping [Lamping 90] and Kathail [Kathail 90] have been proposed that claim to implement optimal λ-reduction. These schemes seem to allow more terms to be shared than are possible using traditional environment or substitution mechanisms. However, they are notable for their extreme complexity, and it is not clear that the overhead incurred by these schemes in order to ensure family classes are always shared is not prohibitive. … [Field 91], p.200.

At this point a few introspective words about this report are in order. Our presentation of Lambda Calculus reduction is rather unconventional because it focuses on practical rather than theoretical economy. Head Order Reduction with sharing can safely be regarded as having many welcome properties and appears to be a good base on which to build more complete reduction systems.

The overall style of our presentation is informal and relies heavily on intuition, liberal use of examples and detailed traces. The Lambda Calculus is a topic that has intrigued many of the world's best mathematicians and has invited deep theoretical treatments. More formal presentations for most of the ideas in this thesis could have been made and the process would probably have shed additional light. Results of corresponding formal treatments are freely used throughout.

A conscious effort is made to include examples which are compact but not trivially simple—so that their inclusion is practical but at the same time illustrative of the complexities involved. The test expression $((\lambda a.a\ a)\ (\lambda bc.\ ((\lambda d.d\ (c\ d))\ ((\lambda e.e\ c\ e)\ b))))$ is repeatedly utilized even though it requires close to sixty reductions under normal order[9]. It is chosen, despite its size, because it exhibits complications and illustrates points that are not evident in situations of contrived simplicity.

---

[9] Plain *HOR* requires more than four hundred iterations. Comparable effort is expended via other normal-order algorithms before the introduction of sharing.

Chapter 2

Syntax, Notation and Representations

2.1 SYNTAX, SEMANTICS AND PRAGMATICS

The Lambda Calculus is a theory emphasizing the computational aspects of functions. It captures in a fundamental manner the concept of producing a result from an argument using a *rule*; i.e., a process having a definition rather than just a static mapping or a set of argument-value pairs. Because of this emphasis, it is well suited in modeling mechanical computation.

Having functions serve as both rules and values—which are full-fledged functions themselves—leads to a type-free structure. In such structures self-application is possible. This generality renders operational techniques developed for Lambda Calculus reduction useful under a wider range of practical settings. The type free nature of the calculus corresponds to the fact that both programs and data are just collections of binary information at the machine level.

In this section the basic elements of this theory are reviewed briefly. For a detailed treatment the reader is referred to the text by H. Barendregt [Barendregt 84] or that by J.R. Hindley and J.P. Seldin [Hindley 86].

Syntax

Two important concepts of functions are reflected in the syntax of the Lambda Calculus. *Application* is the primitive operation that brings together a function with an argument; a function f applied to an argument a is written by a simple juxtaposition, fa. The operation which plays a dual role to application is called *abstraction*. Abstraction is the mechanism by which occurrences of potential arguments within an expression are declared. In the syntax of the modern Lambda Calculus, abstraction is denoted by an abstractor λ, followed by a variable identifier, the *binder*, followed by a delimiting period, which is followed by the *body* of the abstraction. Occurrences of this variable identifier within the body of an abstraction imply a literal substitution of the variable occurrences by any argument to which the abstraction may be applied. For example, if $f(x)$ is a function of x—written in a traditional infix notation—then $\lambda x.f(x)$ denotes the function f itself since it maps an argument, say a, to $f(a)$.

Application of multiple arguments is captured by a sequential application of one argument at a time. For example, if $f_{xy}(x,y)$ is a binary function, one can define $f_x = \lambda y.f_{xy}(x,y)$ and $f = \lambda x.f_x$. Then the following expressions are all equivalent:

$$(fx)y = f_x y = f_{xy}(x,y). \tag{2.1}$$

The syntax for *lambda terms* or simply *terms* is defined next. If x is a variable and e, s, t are terms, then the well-formed expressions (*wfe*s) are given by the rule

$$e := x \mid \lambda x.s \mid st \mid (e). \tag{2.2}$$

Application is left-associative. Parentheses are used to alter the order of application. Nested abstractions, e.g., $(\lambda x.(\lambda y.xyx))$ are usually written with a single abstractor followed by all the binders, followed by a single period, i.e., as $\lambda xy.xyx$.

A variable x occurs *free* in a term if it is not inside the scope of an abstraction with x as its bound variable. It occurs *bound* if it does occur within the scope of such an abstraction. Terms that differ only by having distinct identifiers for their corresponding binders and, possibly, for corresponding bound variable occurrences, are called *α-congruent*. It is said that they result from each other via *α-conversion*. Such terms are commonly identified because this convention simplifies the rules for substitution.

Reduction

Within a lambda expression, the process of repeatedly combining abstractions with the subterms to which they are applied is called *reduction*. Via reduction the structure of an expression is "simplified" in the sense of getting closer to an invariant or *normal* form. The conversion rules for the Lambda Calculus are called the α, β and η-rules. The α-rule entails only the above mentioned change of name of a binder and the of corresponding variable occurrences and therefore does not lead to a simpler expression. The β-rule states that an expression of the form $((\lambda x.F)\ G)$, called a *redex*, can be transformed by substituting G for all free occurrences of x in F. The η-rule captures the principle of *extensionality*; since lambda terms denote processes, terms that produce the same result when applied to any third term are thought of as equivalent. According to the η-rule, $\lambda x.Mx$ and M are identified if x does not occur free in M.

In this dissertation we only need to focus on the β-rule; the notation chosen for our operational methods does not use variable identifiers at all and thus renders α-conversion inessential. The η-rule presents special difficulties with respect to mechanization and is of questionable practical importance. The conversion rules may be applied in any order. The Lambda Calculus is a *confluent* system which means that the reduction sequences permitted by the rules above are guaranteed to yield α-congruent results without regard to the order in which the rules are applied.

Substitution within the body of an abstraction has to be defined carefully in order to avoid the possibility of *capturing* relatively free occurrences of the abstraction's binder by an unrelated but identically named binder. This complication presents many practical difficulties and has to be dealt with effectively if direct Lambda Calculus reduction is to be done efficiently.

Not all lambda expressions have normal forms and not all reduction sequences terminate with a normal form. An expression can have such a form that in the process of β-reduction additional redices are generated, with or without a corresponding increase in size. The classic example is the term $\Delta = (\lambda x.xx)(\lambda x.xx)$ which reduces to itself, ad infinitum.

A *reduction strategy* determines which redex to reduce when more than one is available in an expression. The *leftmost* redex is the one whose λ is to the left of the λ's of all the other redices. Conversely, the *rightmost* redex is the one whose λ is to the right of all other redices in an expression. The *outermost* redex is one not textually contained within any other redex. Similarly, an *innermost* redex is one that does not contain any other redices.

Under an *applicative order* reduction strategy, the redex to be contracted in each step is the leftmost among the innermost ones. Applicative order can save work by ensuring that a needed argument is reduced once, and then used in all needed places. But applicative order is not complete; in its eagerness it may attempt to reduce a subterm that has no normal form but whose reductum is not required for reaching an overall normal form. For example, $(\lambda xy.y)\ \Delta$ has a normal form, namely $\lambda y.y$, since the diverging argument is discarded. But an applicative order reduction of this expression never terminates.

Under a *normal order* reduction strategy, the outermost leftmost redex is always contracted. A normal strategy is normalizing by being ultimately cautious. An operand expression is substituted "as is" for all occurrences of the bound variable. If the bound variable occurs few times or not at all or when the reduction of an operand expression is easy, normal order is a good strategy because it combines efficiency with completeness. If multiple occurrences of the bound variable are present or if the reduction of an operand has a high cost then normal order can be extremely inefficient. Normal order reduction nonetheless is guaranteed to yield a normal form if one exists.

Chapter 3 will make clear that a concrete reduction *algorithm* is more involved than simply choosing a reduction strategy. Even while obeying the same reduction strategy, Lambda Calculus reducers can utilize different representations for expressions, employ distinct traversal mechanisms and various bookkeeping structures. In sort, the operationally interesting and practically significant aspects of reduction are in the details. Normal order, with variations to make it more practical, is the strategy pursued further in this thesis; applicative order is also discussed, primarily as a point of reference.

Normal Forms

An expression that contains no β-redices is in *strong β-normal-form* or simply *normal form* (*nf*). This normal form is viewed as canonically representing the set of all lambda expressions that reduce to it. Operationally, the *nf* is the intuitively desired "result" of a computation. In addition to *nf*, two other, much less final, notions of normal form are important.

An expression is in *weak head normal form* (*whnf*) if it is not an application. In the pure Lambda Calculus, expressions in *whnf* are those that are top-level abstractions (2.3). In Combinatory Logic or in an applied Lambda Calculus which includes strict functions, an expression is also defined to be in *whnf* if it consists of a *partially* applied functor; i.e., one which is applied to fewer than the minimally required arguments (2.4).

$$W_\lambda = \lambda x_1...x_m . B \tag{2.3}$$
$$W_c = F E_1 ... E_n, \text{ where } F \text{ is of arity } k, k > n. \tag{2.4}$$

From a computational viewpoint a *whnf* is an operator which is inactive because it is either partially applied or not applied at all.

A *head normal form* (*hnf*) is an expression having a top-level structure that is impervious to further reductions. It is thus constrained to be of the form

$$H = \lambda x_1...x_m . x_i M_1...M_n \text{ where } 1 \le i \le m. \tag{2.5}$$

The *head variable* of H is x_i. It is easily seen that contractions of any redices within the M_i's cannot affect the top-level structure of H. Head normal forms are important for many reasons. They represent a stage in reduction in which interaction between subterms is bounded. They also represent terms that are meaningful or *solvable*. A term is solvable if it can be reduced to the identity term $I = \lambda x.x$ when applied to suitable arguments. Terms without normal forms can still be solvable; terms that are not solvable are the nonsensical terms of the calculus. For example, the paradoxical combinator $Y = (\lambda x.(\lambda z.zz)(\lambda y.x(yy)))$

reduces to *I* when applied to *KI*, where $K = \lambda xy.x$, even though it does not possess a normal form. On the other hand, $\Delta = (\lambda x.xx)(\lambda x.xx)$ does not have a *hnf* and it is not solvable.

Normal, head and weak normal forms constitute a hierarchy. Every closed expression in *nf* is also in *hnf*. Similarly each expression in *hnf* is also in *whnf*.

Delayed Substitutions

When an abstraction is combined with an argument, the substitution implied by the semantics of a β-reduction may be fully effected at once or it can be delayed. It is usually more efficient to just make a record of the mapping than to completely traverse the body of the abstraction in search of the binder occurrences during each and every reduction step. The mappings are collected and consulted on-demand as variables are encountered during traversal of the body.

This important realization leads us to two new concepts; the *closure* and the *environment*. An environment is a mapping from variables to expressions. More specifically, it is a collection of delayed substitutions. A closure is a pair consisting of an expression, which typically has free variables, together with an environment that gives values to the expression's free variables. With closures and an environment, substitution is no longer one large atomic action but becomes incremental and distributed over time.

These concepts were clarified by Peter Landin [Landin 64] in the context of evaluation of applicative (side-effect free) expressions. Closures are identical to *Lisp*'s FUNARGs, the traditional mechanism employed for implementing function calls under lexical scoping. Closures and, their generalization, suspensions, are in the center of the reduction mechanisms considered[10]. It is the "wholesale" treatment of all pending substitutions that makes environment-based reduction economical.

Giving more examples of reduction is postponed until after the discussion of concrete representation choices and the introduction of a notation that ties everything together.

2.2 ABSTRACT REDUCER REPRESENTATIONS

This section outlines the representation choices for lambda expressions and runtime structures employed in our *abstract* reducers. Throughout this thesis a distilled reduction algorithm for the pure Lambda Calculus is referred to as a abstract reducer—the emphasis of such algorithms being the fundamental issues of reduction. These reducers are actually very

---

[10] The names closure, suspension or recipe are commonly employed interchangeably. In this thesis the name *closure* is reserved strictly for an abstraction-environment pairing and the name *suspension* for the more general case of any term-environment pairing. Under weak reduction without sharing only pure closures are formed.

concrete in the sense of relying on concrete representations and having a concrete realization. An *applied reduction system* on the other hand refers to more detailed implementations of enriched Lambda Calculus systems like those of *BETAR3*, *HORSE* and *π-RED* (see Appendices A and C).

Because abstract reducers are limited to the pure calculus, the necessary structures are simple and few. Our representation choices are general enough so that they can be used with both major techniques of variable codings. The ordinary representation choices for Head Order Reduction are referred to collectively as the "HOR-style" and those of the Wadsworth/Aiello–Prini are referred to as the "REP-style"[11]. In choosing representations we opted for generality rather than absolute minimality of structure in order to make comparisons during the evaluation phase easier and more fair.

Lambda Expression Representations

Table 2.1 summarizes several styles of displaying lambda expressions. The expression $(\lambda xy.x)(\lambda y.y)$ is depicted in eight ways as an example of our notational and representational choices. After the classic linear and tree notation of row (*A*), we list in row (*B*) the expression in *BTF*—a minimal number-based notation described in the next section. In row (*C*) each application, abstraction, and variable node is fully tagged for identification. This is the uniform concrete structure utilized by all the abstract reducers. It is presented in detail below. Finally, row (*D*) shows a linear representation with forwarding pointers that relies on the adjacency of memory locations to silently emphasize the left spine. This is the preferred representation for a low-level machine implementation; it is discussed in section 8.1.

Generic Structural Tagging

Lambda expressions and runtime structures are represented by graphs. Each possible type of node at the calculus level is mapped to a list at the concrete machine level. The first element of each such list is a tag that signifies the type of the node. Accessor macros only need to check this tag to determine the type of the node during traversal.

For example, an application node is simply a three element list with a tag and two pointers to the function and argument:

```
•---•---•---nil
|   |   |
'app fun arg
```

---

[11] REP-style is borrowed from the representation described by Aiello and Prini [Aiello 81]. "*REP*" is the name of the function that transforms a lambda expression to this representation.

	Name	Linear	2-D

A.	Classic	$(\lambda xy.x)(\lambda y.y)$	
			
			``
			                    @
			                 / \
			                / λ
			               / /
				\
				              λ λ y y
				            / \
					
					           x λ y
					
						\
						                y x
						``

B.	BTF	$((-2\ 1)(-1\ 0))$	``` @---\---0 |
1 	 \---1 	 2 ```

C.	Structural Tagging	```
		(app
		 (lam (var _) (lam (var _) 1))
		 (lam (var _) 0))
		``
			``
			•---•---•---nil
			| |
			app •---•---•---nil
			
					|
					    lam
						0
						        •---•---nil
						
							|
							        var _
							•---•---•---nil
							| |
							lam •---•---•----nil
							
									|
									    lam
										1
										        •---•---nil
										
											|
											        var _
											•---•---nil
											| |
											var _
											``

D.	Linear Left Child (CAR-coding) with variable names	``` (APP-PTR (LAMBDA X):(LAMBDA Y): (VAR 1) (LAMBDA Y):(VAR 0)) ```	``` 	AP	•
LAM	X
LAM	Y
H VAR	1
LAM	Y
H VAR	0	 ```

Table 2.1 Several depictions of the lambda expression $(\lambda xy.x)(\lambda y.y)$.

For readers familiar with *Lisp* and in order to clarify the meaning of the macros appearing in the definitions of Appendix A, the constructor macro *mk-app* is written as

(defmacro mk-app (fun arg) `(list 'app ,fun ,arg)).

Similarly, an abstraction node includes pointers to the binder and the body, or *form*, of the abstraction:

```
•---•---•---nil
|   |   |
'lam bnd frm
```

The binder node is not essential for reduction in HOR-style representations; its sole utility is to provide a place to store a user-level identifier for the variable under question. The binder nodes though, are an essential element of the REP-style because they are the target of the pointers emanating from the variable occurrences of the binder within the body of the abstraction:

```
•---•---nil
|   |
'var nam
```

In the HOR-style representations, the variable occurrences are represented by non-negative integers according to the *de Bruijn* (*dB*) convention [Bruijn 72]. According to this convention, an occurrence of the binder variable within the body of an abstraction is represented by a non-negative integer that counts the number of any intermediate abstractions[12]. For example, $\lambda xy.xyx$ is written as $\lambda\lambda.101$.

A predicate, *is-app*, which returns true when its argument is an application is

(defmacro is-app (term) `(eq (car ,term) 'app)).

The corresponding accessor macros, *fun-of* and *arg-of* are entirely straightforward.

Suspensions and Linking of Environment Elements

A suspension node includes pointers to the term and environment portions of the suspension. Closures and suspensions are represented identically save for the appropriately named tags appearing as the first element.

```
•----•----•---nil
|    |    |
'susp term env
```

---

[12] The binding indices in this work start counting at zero, a practice which is beneficial when they are employed to index directly into environment structures. Other researchers, following de Bruijn, have settled on conventions that have the indices starting at one.

The environment structure itself takes two distinct forms depending on the representation family of the abstract reducer. In the HOR-style the environment is a list of values. Variables, represented by their integer de Bruijn indices, are looked-up by direct indexing. There is no need for any additional information. The ith element of the environment list holds the value of the variable with index i.

```
•----•---..--env
|    |
val0 val1
```

In the REP-style the environment is a list of pairs. Each pair has a variable and a value element. Variables, represented by pointers to their corresponding binders, are looked up by sequential comparison starting from the "tip" or last entry of the environment.

```
•---------•---..---env
|         |
•---val0  •---val1
|         |
var0      var1
```

Figure 2.1 is an example of all these structures at work. It shows the representation of the *hnf* for the expression *tw7 = (λbc. ((λd.d (c d)) ((λe.e c e) b)))* as produced via HOR reduction. This special intermediate form is explained in section 4.1. For the moment, the reader is asked to concentrate only on its structure.

The two suspensions of this *hnf* are highlighted. Suspension *S1* has a single application as its term and has a three-element environment. Suspension *S2* itself constitutes the first element of that environment. Suspension *S2* has an environment of two entries. The meaning of such environment entries consisting of negative integers is explained in section 3.6. The tree of Figure 2.1 (*A*) is an unraveling of the graph structure shown in (*B*); the last two elements of the environment portions of *S1* and *S2* in reality are shared.

An Example of two Representations

To further demonstrate the range of representation choices we present one last example. One of the simplest lambda expression with an interesting, in the sense of not being trivial, structure is the *Y* combinator. Figure 2.2 part (*A*) and part (*B*) shows *Y* in the two representation styles that we have described. For the sake of comparison, part (*D*) shows a minimal "raw" representation which is indeed adequate for reduction.

The structural tagging conventions used in the abstract reducers do not constitute by far the most economical representation— at least not for HOR. The "raw" one is much more compact. The reader should keep in mind that all the numbers for memory consumption reported in the evaluation chapter of this thesis are with respect to the general but non-optimal structure-tagging representations.

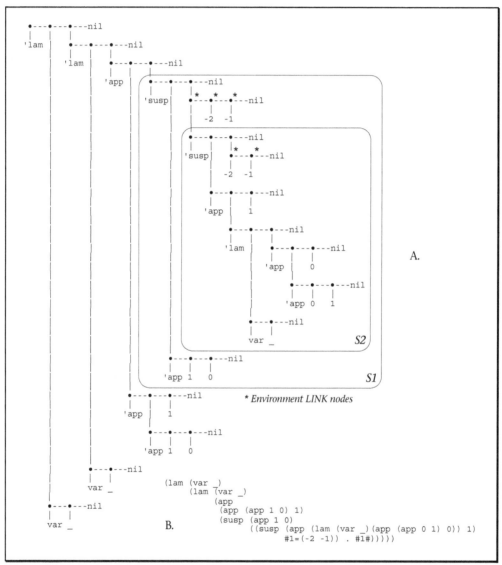

Figure 2.1 Representation of hnf of test expression tw7.

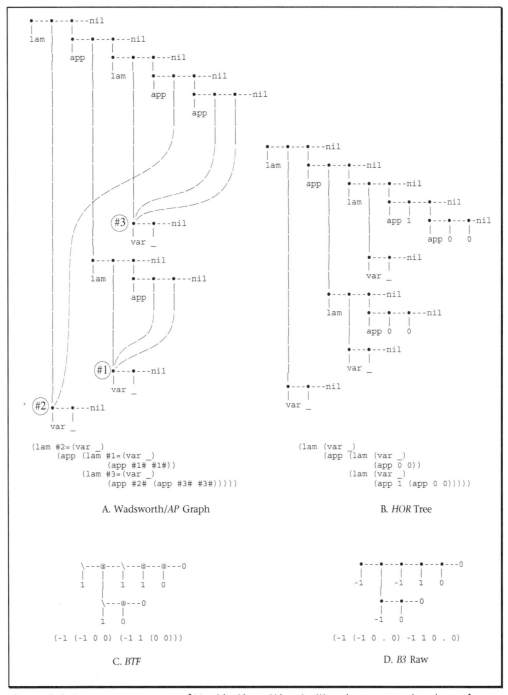

Figure 2.2 Four representations of Y = (λx.(λz.zz)(λy.x(yy))), with corresponding linear forms.

2.3 BTF - A GENERIC OPERATIONAL NOTATION

While studying reduction in a de Bruijn (*dB*) notation one quickly recognizes the need for inventing a minimal notation reflecting the structure of an expression. A notation of this sort should visually emphasize the differences of nearly identical expressions and, more importantly, it should have the property that it may be used as a syntactic normal form. It would also be useful if such a notation can be extended to include structures of a more operational nature that are present *during* graph reduction.

Syntax for Lambda Expressions

An intermediate form which satisfies the former demands is given the name *BTF* (from the initials of the originator and this author). It is a natural extension of the raw "integer form with parentheses" utilized in some of the example expression in the first description of Head Order Reduction [Berkling 87]. Using this notation, lambda expressions are given compact linear descriptions[13].

Its syntax is as follows:

$$
\begin{aligned}
Term &:= Abstr \mid AppList \mid Var \mid (Term) \\
AppList &:= \varnothing \mid (Term \cdot AppList) \\
Abstr &:= (Lam \cdot Term) \\
Var &:= Nat \\
Lam &:= -Nat \\
Nat &:= 0 \mid 1 \mid \ldots
\end{aligned}
\qquad (2.6)
$$

This choice for a syntax is a natural one once numerical offsets for bindings are adopted. One would like applications to be denoted simply by textual juxtaposition and a sequence of abstractions by its only significant property, the number of lambdas. According to this syntax, the expression $\lambda xy.y$ which in de Bruijn notation is ($\lambda\lambda$. 0) becomes (-2 0) and $\lambda xyz.xyz(yz)$ which is ($\lambda\lambda\lambda$ 2 1 0 (1 0)) becomes (-3 2 1 0 (1 0)). They are summarized in Table 2.2. The fact that the number of nested abstractions is represented by a negative integer is of significance; under this convention m and k in the term $(\ldots k \ldots (-n \ m) \ldots)$ refer to the same binder iff $k = m - n, m \geq n$.

---

[13] All the well-formed expressions (*wfe*'s) in our treatment are closed and, hence, are either abstractions with an ample number of binders, or combinations of similarly closed *wfe*'s. This assumption is not restrictive from a practical perspective; globally free variables can be regarded as referring to globally distinct lambda binders. This can be achieved by assigning to each such variable a *dB* index of sufficient magnitude. See [Bruijn 72].

Classic	de Bruijn	BTF
$\lambda xy.\,y$	$(\lambda\!\!\lambda\ .0)$	$(\text{-}20)$
$\lambda xyz.\,xyz(yz)$	$(\lambda\!\!\lambda\!\!\lambda\ .210(10))$	$(\text{-}3210(10))$
$(\lambda h.(\lambda w.\,ww)(\lambda x.\,h(\lambda y.\,xxy)))$	$(\lambda(\lambda 00)(\lambda 1(\lambda 110)))$	$(\text{-}1(\text{-}100)(\text{-}11(\text{-}1110)))$

Table 2.2 Lambda expressions in classic, de Bruijn and BTF notation.

The reader should not be left with the impression that *BTF* is proposed for human consumption and everyday use. Variables, with their mnemonic identifiers, serve infinitely better in the place-holding role. But when reduction issues are discussed, especially those of operational nature, variable names are not to be missed because they complicate treatment substantially. After familiarization, one discovers that de Bruijn indices and the lack of extra decorations is not at all a burden.

In contrast with combinator expressions that result from an abstraction procedure, the structure of an expression in *BTF* is still highly visible and has a trivial correspondence to the original. With some practice, the rules of reduction can be applied with ease. But the real benefit of *BTF* is indeed the facilitation of the exploration of the various reduction schemes. It provides us with a compact notation, explicitly designed for reduction.

Simplification Rules

The syntax rules (2.6) of *BTF* allow for ambiguity in the linear representation of the terms—analogous to that of S-expressions denoting lists. The ambiguity is removed if the standard simplifications (2.7) are performed. As a consequence, a unique *BTF* for every expression can be reached.

$$
\begin{aligned}
((\mathit{Expr})) &\Leftrightarrow \mathit{Expr}) \\
(\mathit{Expr}1\,(\ \mathit{Expr}2)) &\Leftrightarrow \mathit{Expr}1\ \mathit{Expr}2) \\
(\mathit{Lam}1\,(\ \mathit{Lam}2\cdot\mathit{Expr})) &\Leftrightarrow [\![\mathit{Lam}1+\mathit{Lam}2]\!]\cdot\mathit{Expr})
\end{aligned}
\tag{2.7}
$$

According to the last set of rules, the term (-6 ((3 4) (0 (-3 (-2 1 0))))) should be simplified to (-6 3 4 (0 (-5 1 0))), a syntactic normal form. The double square brackets in the third rule of (2.7) denote integer addition of the component indices.

Reduction Example

Let us now proceed with a reduction example. The expression $(K'\ \underline{1}\ \underline{2}\ \underline{3})$ is chosen where K' is the combinator $(\lambda xy.y)$ and the underlined numerals are expressions consisting of the Church representation of integers. Mutual application of Church Numerals amounts to exponentiation (in the opposite order) and a result of $(\underline{2}\ \underline{3})$ or $\underline{3^2}$ or $\underline{9}$ is expected. The test expression in *BTF* is $e = ((\text{-}2\ 0)(\text{-}2\ 1\ 0)(\text{-}2\ 1\ (1\ 0))(\text{-}2\ 1\ (1\ (1\ 0))))$.

A two dimensional tree notation, again due to Berkling, is also utilized. It is a rotated (45°, counterclockwise) view of the more common depictions of lambda trees. It uses few extra symbols in addition to the integer indices and it has the advantage that it can be generated mechanically (See Appendix C for details on our experimentation system). The lambda expression *e* above in this notation is the tree shown as the initial step of Figure 2.3.

Initially, reduction in *BTF* is easier to comprehend if one concentrates on the tree sequence of Figure 2.3 rather than the linear sequence of Table 2.3 below. The left spine of a term in tree form is the sequence of the leftmost vertical lines—including the topmost abstractions, if any—plus a last element which is always a variable. This variable at the end of the left spine is the *head* of the expression (cf. equation 2.5 of section 2.1). Under a normal order strategy, the redex to be contracted in each step is the first occurrence of an application symbol (@) directly over a lambda symbol (\) as the tree is traversed along the

0.	((-2 0) (-2 1 0)(-2 1 (1 0))(-2 1 (1 (1 0))))	*e*
1.	((-1 0) (-2 1 (1 0))(-2 1 (1 (1 0))))	
2.	((-2 1 (1 0)) (-2 1 (1 (1 0))))	
3.	(-1 (-2 1 (1 (1 0))) ((-2 1 (1 (1 0))) 0))	*whnf*
4.	(-2 (-2 1 (1 (1 0))) 1 ((-2 1 (1 (1 0))) 1 ((-2 1 (1 (1 0))) 1 0)))	
5.	(-2 (-1 2 (2 (2 0))) ((-2 1 (1 (1 0))) 1 ((-2 1 (1 (1 0))) 1 0)))	
6.	(-2 1 (1 (1 ((-2 1 (1 (1 0))) 1 ((-2 1 (1 (1 0))) 1 0)))))	*hnf*
7.	(-2 1 (1 (1 ((-1 2 (2 (2 0))) ((-2 1 (1 (1 0))) 1 0)))))	
8.	(-2 1 (1 (1 (1 (1 ((-2 1 (1 (1 0))) 1 0)))))))	
9.	(-2 1 (1 (1 (1 (1 ((-1 2 (2 (2 0))) 0)))))))	
10.	(-2 1 (1 (1 (1 (1 (1 (1 (1 0)))))))))	*nf*

Table 2.3 Normal-order reduction of ((-2 0)(-2 1 0)(-2 1 (1 0))(-2 1 (1 (1 0)))).

left spine. A closed expression can be substituted directly for all occurrences without undergoing any changes (cf. steps 2 → 3). An expression with relatively-free variables must have such, and only such, variables adjusted when a substitution is made inside an abstraction (cf. steps 3 → 4). An expression is in normal form when no application node (@) is directly above an abstraction node (\) (cf. step 10).

In Table 2.3 the same reduction sequence is depicted in the linear *BTF* format with the redex of each step underlined and the operator and operand separated by long spaces. This example is chosen partly because it has distinct weak, head and strong normal forms.

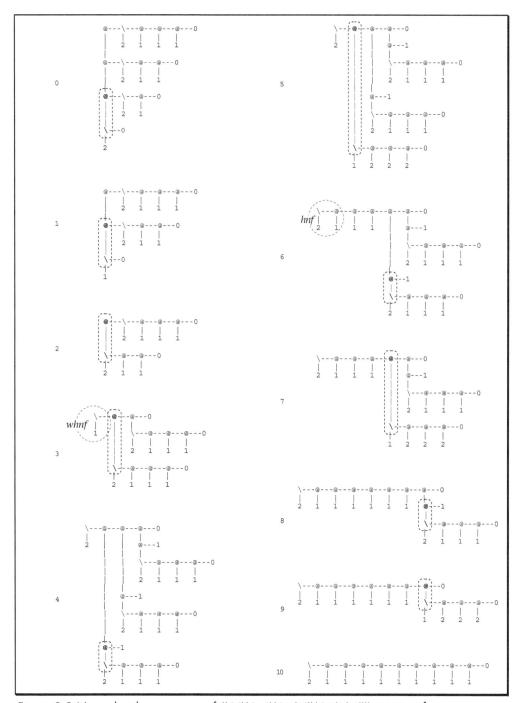

Figure 2.3 Normal order sequence of ((-2 0)(-2 1 0)(-2 1 (1 0))(-2 1 (1 (1 0))))   in tree form.

Environments and Suspensions

The rest of this section expands the *BTF* notation by adding operationally-oriented features. Some of these features are general and others are specific to HOR. This, and the next two subsections, may be skipped during a first reading. They appear here in order to retain the complete description of the *BTF* syntax in a single section.

First, *BTF* is extended to encompass closures/suspensions and environments. The syntactic convention for the representation of a closure or a suspension is meant to visually emphasize the pairing of a, typically open, term with an environment.

$$[(\ldots)$$
$$\{ \ldots \ldots \ldots \}]$$

Term   Environment

By virtue of the de Bruijn representation of variables the correct binding for a relatively free variable occurrence in a closure term is easy to find: it is simply the result of a direct indexing operation into the environment structure. For example, the closure expression

$$[(-1 \boxed{1} (1\ 0))$$
$$\{ [(-2\ 1\ (1\ (1\ 0))) \{ \}] \}]$$

denotes that the leftmost variable of the term component with a de Bruijn index of 1 (shown boxed) is bound to the closure expression [(-2 1 (1 (1 0))){ }] which occupies position zero in the environment sequence. This binding is the appropriate one since the boxed variable "reaches over" one element deep (i.e., into position 0) into the environment after being adjusted to counter the effect of its surrounding one lambda. In general, the environment will contain a list of suspensions many of which will have multiple relatively free variables and, hence, multiple, recursive lookups will be needed to completely determine the meaning of a an open term. The term above is the final result of a *whnf* reduction of the expression $e = ((-2\ 0)\ (-2\ 1\ 0)\ (-2\ 1\ (1\ 0))\ (-2\ 1\ (1\ (1\ 0))))$ (cf. with the result of step 3 of the normal order reduction in Table 2.3). The environment in the case of an operational weak head normal form consists always of a list of closures.

In addition to being an operationally oriented tool, this notation has close correspondence to independent theories which formalize the notion of substitution. A more precise definition of closure expressions such the above is given in section 3.6 which describes the meaning and shows a graphic depiction of such expressions.

The extended *BTF* has the following additional rewrite rules that make concrete the introduction of suspensions and environments as first class elements.

$$Susp \; := \; [\,Term \; \{\; EnvList \;\}\,]$$
$$EnvList \; := \; \varnothing \mid (EnvEl \cdot EnvList)$$
$$EnvEl \; := \; Susp$$

(2.8)

Unbound Variable Counts

Strong HOR reduction, which advances beyond *whnf*s to *hnf*s and *nf*'s, suggests the next extension to our linear notation. Environments can have a new sort of element which is inextricably related to HOR reduction utilizing de Bruijn indices. A possible entry may be the count of the number of times the composite substitutions, represented by the state of the environment, have been "carried" over an unapplied abstraction. Such elements are named *unbound variable counts* (*UVC*s); they are usually referred to by the symbol ϕ and take the form of a negative integer. This extension of *BTF* is defined by a modification of the *EnvEl* rule of (2.8):

$$EnvEl \; := \; Susp \mid UVC$$
$$UVC \; := \; -Nat$$

(2.9)

The meaning of the last extension will only become clear after the description of the *UVC* mechanism in detail in section 4.2. We proceed with an example anyway.

Consider the operational *hnf H* shown as the following graph[14],

$H =$ (lam 1 (app (app (app (app 0 #3=(susp (lam 2 (app (lam 1 (app 0 #1=(app 1 0))) #2=(app (lam 1 (app (app 0 1) 0)) 1))) nil)) 0)(susp #1# (link (susp #2# #4=(link #3# (link -1 nil))) #4#))) (susp #1# (link (susp #2# #5=(link -1 (link #3# nil))) #5#))))[15]

Unraveling *H* produces the following *BTF* expression:

$B =$ (-1 0 [(-2 (-1 0 (1 0)) ((-1 0 1 0) 1)) { }] 0 [(1 0) { [((-1 0 1 0) 1) { [(-2 (-1 0 (1 0)) ((-1 0 1 0) 1)) { }] -1 }] [(-2 (-1 0 (1 0)) ((-1 0 1 0) 1)) { }] -1 }] [(1 0) { [((-1 0 1 0) 1) { -1 [(-2 (-1 0 (1 0)) ((-1 0 1 0) 1)) { }] }] -1 [(-2 (-1 0 (1 0)) ((-1 0 1 0) 1)) { }] }]).

Occurrences of *UVC*s in the environment subterms of *H* and *B* are doubly underlined. There are two occurrences of *UVC*s in graph *H*; through unraveling, they result in four occurrences in expression *B*.

---

[14] Throughout, we consider head normal forms that result from the reduction of a *closed* initial expression. The previous comment about variables of a global nature still applies.

[15] Listed here only for purposes of illustration, graph *H* is the result of head-normal reduction of the *wfe* ((-1 0 0) (-2 (-1 0 (1 0)) ((-1 0 1 0) 1))). Algorithm *HN-HOR* is used. Appendix B has a detailed trace of the required steps (Table B.9). A graphical depiction appears in section 4.1 (Figures 4.1 & 4.2).

Abstraction-Depth Context Tagging

Extension (2.10) of *BTF* is useful when HOR is augmented with general sharing. It reflects the ability of having shared *nf*'s as intermediate results of reduction as explained in detail in section 6.3. The enhancement to *BTF* consists of the addition of tagged *nf*'s to the set of possible element types of an environment entry. The tag signifies the prevailing number of abstractions making up the outermost part of the operational *hnf* that was in effect when the tagged *nf* was produced. Such tagging permits an artificial closing of an expression which is in *nf* but which may have relatively free variables "out of sync" with its final context. Tagging via a pseudo floating-point number is an encoding trick that maintains the number-based theme of *BTF*.

$$
\begin{aligned}
EnvEl &:= \quad Susp \mid UVC \mid ShExp \\
ShExp &:= \quad (\phi.0 \ Term_{nf}) \\
\phi &:= \quad Nat
\end{aligned}
\qquad (2.10)
$$

Again, we proceed to show an example expression while postponing further semantic details:

$$T = [\ (1 \ 0 \ 1) \ \{ \ (\underline{\underline{1.0}} \ \text{-2} \ 1 \ 0 \ 1 \ (0 \ (1 \ 0 \ 1))) \ \text{-1} \ \text{-1} \ \}]$$

The environment entry at position 0 of the suspension term T is a *nf* tagged with a creation context that consists of a single abstraction. In this specific case the *nf*, (-2 1 0 1 (0 (1 0 1))), is closed so the abstraction-depth context is immaterial.

BTF Syntax Summary

Recapitulating, we have presented a notation which exposes the operational aspects of reduction with the de Bruijn convention for variables. It is our belief that *BTF* itself may be viewed as a combinatory logic; see for example Curien's text [Curien 86].

For easy reference, the complete syntax of *BTF* appears in (2.11). The first six rules cover pure lambda expressions and the last six cover the additional entities that arise in HOR and related reduction. The simplification rules (2.7) still apply.

$$
\begin{aligned}
Term &:= \quad Abstr \mid AppList \mid Var \mid (Term) \\
AppList &:= \quad \varnothing \mid (Term \cdot AppList) \\
Abstr &:= \quad (Lam \cdot Term) \\
Var &:= \quad Nat \\
Lam &:= \quad -Nat \\
Nat &:= \quad 0 \mid 1 \mid \ldots
\end{aligned}
\qquad (2.11a)
$$

$$
\begin{aligned}
Susp &:= [Term \quad \{ \quad EnvList \quad \}] \\
EnvList &:= \varnothing \,|\, (EnvEl \cdot EnvList) \\
EnvEl &:= Susp \,|\, UVC \,|\, ShExp \\
ShExp &:= (\phi.0 \quad Term_{nf}) \\
UVC &:= -Nat \\
\phi &:= Nat
\end{aligned}
\qquad (2.11b)
$$

If mixed terms are admitted (cf. section 4.5 and chapter 5) then the rule for *Term* must be augmented to allow *Susp*s.

For a final illustration of *BTF*, Table B.11 in Appendix B groups together the graph, *BTF* and left spine formats of the operational *hnf* shown previously (expressions *H* and *B*). From this point on lambda expressions are shown mainly in *BTF*.

2.4 TRANSLATIONS TO BTF

BTF is by origin and design closely related to the representations chosen for developing and coding our abstract reducers. The correspondence is formalized (in the " $\rightarrow$ *BTF*" direction only) in this section. It takes the form of two functions *h2b* and *r2b* mapping representation constructs to linear *BTF* terms. The former is used in the families of reducers that employ de Bruijn variables and unbound variable counts (*UVCs*) and the latter in those families that rely on pointer variables in the example of Wadsworth and Aiello–Prini.

$$
\begin{aligned}
&h2b \quad nil = nil \\
&h2b \quad JM = {}'@ \\
&h2b \quad LM = {}'\backslash \\[4pt]
&h2b \quad VAR\,n = {}'n \\
&h2b \quad UVC\,\phi = {}'\phi \\[4pt]
&h2b \quad LAM\,a = {}'('{-}1\ '(\ h2b\ a\ ')') \\
&h2b \quad APP\,a\,b = {}'(\ h2b\ a\ h2b\ b\ ') \\
&h2b \quad SUSP\,t\,e = {}'[\ h2b\ t\ '\{\ h2bl\ e\ '\}'] \\
&h2b \quad SHEXP\,a\,\phi = {}'(\ '{-}\phi.0\ h2b\ a\ ') \\[4pt]
&h2bl \quad nil = nil \\
&h2bl \quad LINK\,e\,es = {}'(\ h2b\ e\ '(\ h2bl\ es\ ')\ ')
\end{aligned}
$$

Figure 2.4 The translation function h2b from HOR graph representations to BTF.

h2b: HOR to BTF

The function *h2b* which translates from the abstract representation employed in the HOR reducers to *BTF* is shown in Figure 2.4. It deals with terms, closures and potentially shared *nf*'s tagged with a prevailing ϕ. Its domain can also be extended to include the two markers that are present in the *HOR* stack during upward traversal. Thereby complete traces of all the runtime structures of the reducers are given in the linear *BTF*.

There is a close relation of the HOR-style representation to the *BTF* syntax a fact which is reflected in the nearly one-to-one mapping of structural elements to clauses of *h2b*. We do make use of an auxiliary definition, *h2bl* which maps *h2b* over an environment.

r2b: REP to BTF

The function *r2b* translates from the REP-style representations to *BTF*. The reason we would like to produce *BTF* terms from representations that employ pointer variables is because *BTF* is a much more readable notation. The two translation functions are similar. The pointer variables do give rise to a more complicated translation because *BTF* was defined with de Bruijn indices for variables. Independently of this extra complication, we argue later that the avoidance of pointer variables is mostly a positive aspect.

Since the Wadsworth-style binder-pointer representation of variables relies on location rather than abstract indexing, the translation function *r2b* requires one more input

$$lu \quad v \quad \varnothing \quad i = \begin{cases} \downarrow, \quad whnf \\ id(v), \quad nf \end{cases}$$

$$lu \quad v \quad v \cdot c \bullet \rho \quad i = i$$

$$lu \quad v \quad u \cdot c \bullet \rho \quad i = lu \quad v \quad \rho \quad (i+1)$$

$$r2b \quad nil \quad repenv = nil$$

$$r2b \quad VARPTR \ v \quad repenv = lu \quad v \quad repenv \quad 0$$

$$r2b \quad LAM \ v \ b \quad repenv = {}'(\ '-1 \ '(\ r2b \ b \ v \cdot LM \bullet repenv \ ')')$$

$$r2b \quad APP \ a \ b \quad repenv = {}'(\ r2b \ a \ env \ r2b \ b \ repenv \ ')$$

$$r2b \quad SUSP \ t \ \rho \quad repenv = {}'[\ r2b \ t \ \rho ++ repenv \ '\{ \ r2bl \ \rho \ repenv \ '\}']$$

$$r2bl \quad nil \quad repenv = nil$$

$$r2bl \quad v \cdot c \bullet \rho \quad repenv = {}'(\ r2b \ c \ repenv \ '(\ r2bl \ \rho \ repenv \ ')')$$

Figure 2.5 The translation function r2b from REP-style graph representations to BTF.

parameter; a representation environment *repenv*, which is initially empty. The proper *repenv* at each point in the recursive application of *r2b* is needed to map pointer variable occurrences to *BTF* de Bruijn indices (Figure 2.5).

Let us focus now on three interesting aspects of *r2b*. First, a lookup process, *lu*, has to be performed when a variable is encountered. The *repenv* is scanned for an entry that includes the identical (*EQ*) binder-pointer corresponding to the variable. Since the binder-pointer representation of variables is employed for both weak and strong reducers, the proper action in the case of an absent environment entry depends on the context; under a *whnf* strategy it is an error for a look-up to fail. Under a strong strategy like *AP*, an absent entry corresponds to a place-holding fresh variable introduced when an unapplied abstraction is traversed. Such variables can be tagged for tracing purposes with a unique identification. This is precisely what is implied by our presentation of the *AP* family (section 3.5) and the first case of the rule for *lu* reflects this decision.

Second, in the abstraction rule of *r2b*, the body of the abstraction is to be processed in an environment which is the representation environment augmented by a dummy binding consisting of the abstraction's binder. Since no value is associated with this binder any symbol would suffice. For consistency with *HOR*, and in recognition of the similar role played, in our implementations a *lambda-marker* (*LM*) is supplied. Lastly, in the rule for a suspension, the local environment of the suspension must be prepended to the representation environment in effect at that point in the translation.

Chapter 3

Graph Reduction Algorithms

3.1 INTRODUCTION

The term *string reduction* is used to describe algorithms that operate on linear representations for lambda expressions, typically employing one or more stacks for traversal. In contrast, the term *graph reduction* is typically refers to algorithms which (*a*) employ a graph representation for lambda expressions and (*b*) effect substitution fully in each reduction step.

Graphs are the natural representation of lambda terms. It is indeed overly limiting to forgo the random access capability permitted by graph representations and only focus on the tops of one or more stacks. But requirement (*b*) above is too strong. Its popularity is attributed to the importance and strong influence of Wadsworth's technique. In this chapter, as the sequence of algorithms leading up to HOR is presented, we indicate that it is unnecessary for a reduction algorithm to effect substitutions literally to be considered true graph reduction.

This chapter begins with an account of the classic *eval-apply* algorithm and the applicative evaluator provided by the *SECD* machine (cf. Table 1.6). Then we turn to strong graph reduction and present Wadsworth's technique and its delayed substitution evolution by Aiello–Prini. We conclude by focusing on the left spine and the operational *hnf* that are the driving forces behind HOR. Throughout, *BTF* is used to demonstrate the details of the algorithms in a uniform manner.

3.2 MCCARTHY'S EVAL -APPLY

The starting point of our exposition of weak reduction strategies within our uniform notational framework is the *eval-apply* interpreter of John McCarthy [McCarthy 60], [McCarthy 65]. It was devised as a definitional interpreter for the applicative language *Lisp* which was emerging in the early 1960's. This interpreter is an ideal means to introduce the operational details of delayed substitution for Lambda Calculus reduction.

The essence of the interpreter is captured by the rules 3.1. The first three rules define the result of *eval* on variables, applications and abstractions, in that order. Note that *eval* is constrained by the third rule and the single rule for apply to return a closure. There is only one rule for *apply* since he operator and the operand of an *apply* are always closures. The procedure *lu* returns the value of a variable in the current environment.

$$
\begin{aligned}
ev \ x \ \rho &\Rightarrow lu \ x \ \rho \\
ev \ ab \ \rho &\Rightarrow ap \ (ev \ a \ \rho) \ (ev \ b \ \rho) \\
ev \ \lambda x.a \ \rho &\Rightarrow a_x\{\rho\} \\
\\
ap \ a_x\{\rho\} \ c &\Rightarrow ev \ a \ x \cdot c \bullet \rho \\
\\
lu \ x \ \varnothing &= \downarrow \\
lu \ x \ x \cdot a \bullet \rho &= a \\
lu \ x \ y \cdot a \bullet \rho &= lu \ x \ \rho
\end{aligned}
$$

(3.1)

The procedure *eval* turns a function application, commonly called a *combination*, to its weak normal form. An environment ρ is used to record the bindings of the bound variable in a β-redex. It is the quintessential example of the applicative or *eager* evaluation strategy—the presence of the *eval*'s around the operator and operand in the rule for applications ensures that only (weakly) evaluated arguments are ever substituted in the function body. In addition to the environment, the other important concept embodied in this interpreter is the notion of a closure. When an abstraction $\lambda x.a$ is encountered, a package $a_x\{\rho\}$, is made of the abstraction's body together with the binder and the environment which is in effect at the point of encounter.

Table 3.1 shows the *whnf* reduction of the expression $e = ((-2\ 1\ 1)\ (-2\ 1\ 0)\ (-1\ 0))$ via this algorithm. After three reductions, *eval-apply* returns the *BTF* expression [(-1 1 0) {[(-2 1 0) { }]}] which represents, in a lazy manner, the overall *whnf* (-1 (-2 1 0) 0). Such terms are referred to as *operational weak head normal forms* (*owhnf*'s).

This is the first instance of using the notation of the previous chapter for displaying the *trace* of a reduction algorithm. The arguments of *eval* and *apply* are all structures that can be described by *BTF* [16]. In the following, we will sometimes show partial traces and refer the reader to Appendix B for the complete picture.

step	term / rator	env / rand
E0	((-2 1 1) (-2 1 0) (-1 0))	nil
E1	((-2 1 1) (-2 1 0))	nil
E2	(-2 1 1)	nil
R2	([(-2 1 1) ({ })])	
E3	(-2 1 0)	nil
R3	([(-2 1 0) ({ })])	
A4	([(-2 1 1) ({ })])	([(-2 1 0) ({ })])
•E5	(-1 1 1)	((([(-2 1 0) ({ })])))
R5	([(-1 1 1) ({ ([(-2 1 0) ({ })]) })])	
R4	([(-1 1 1) ({ ([(-2 1 0) ({ })]) })])	
R1	([(-1 1 1) ({ ([(-2 1 0) ({ })]) })])	
E6	(-1 0)	nil
R6	([(-1 0) ({ })])	
A7	([(-1 1 1) ({ ([(-2 1 0) ({ })]) })])	([(-1 0) ({ })])
•E8	(1 1)	((([(-1 0) ({ })]) ([(-2 1 0) ({ })])))
E9	1	((([(-1 0) ({ })]) ([(-2 1 0) ({ })])))
R9	([(-2 1 0) ({ })])	
E10	1	((([(-1 0) ({ })]) ([(-2 1 0) ({ })])))
R10	([(-2 1 0) ({ })])	
A11	([(-2 1 0) ({ })])	([(-2 1 0) ({ })])
•E12	(-1 1 0)	((([(-2 1 0) ({ })])))
R12	([(-1 1 0) ({ ([(-2 1 0) ({ })]) })])	
R11	([(-1 1 0) ({ ([(-2 1 0) ({ })]) })])	
R8	([(-1 1 0) ({ ([(-2 1 0) ({ })]) })])	
R7	([(-1 1 0) ({ ([(-2 1 0) ({ })]) })])	
R0	([(-1 1 0) ({ ([(-2 1 0) ({ })]) })])	

Table 3.1 EV-AP reduction of ((-2 1 1)(-2 1 0)(-1 0)).

The first column of Table 3.1 tags the invocations of the rules for *eval* or *apply*. Each invocation or *call* is given a sequence number and it is marked with one of E, A, or R; the latter tags a closure result returned by a rule invocation. Calls that are the result of a beta reduction are preceded by a solid dot. To conserve space, only two columns are used; they hold the term and the environment (in the case of *eval*) and the operator and the operand (in the case of *apply*).

Looking closely at the trace above we now highlight a couple of important cases. An abstraction is simply returned as is; it only needs to be dressed up as a closure with the current environment.

E2:	(-2 1 1)	nil
R2:	([(-2 1 1) ({ })])	

---

[16] There is a slight deviation from the syntax given by the rules (2.11). The bracketing fences of both closures/suspensions and environments are enclosed in an extra set of parentheses.

An operator closure applied to an operand closure always implies a reduction. The reductum is represented by a term which consists of the body of the operator closure together with an environment which is that of the operator closure augmented by the operand closure:

A7: ([(-1 1 1) ({ ([(-2 1 0) ({ })]) })])   ([(-1 0) ({ })])
•E8: (1 1)   ((([(-1 0) ({ })]) ([(-2 1 0) ({ })]))

When no *eval* calls are pending the evaluation is complete.

Let us now informally follow an eager strategy that caters to multiple arguments and does not use a delaying mechanism[17] (Table 3.2). First, one would have to reduce the two operand expressions to their respective *whnfs*. In this case, both (-2 1 0) and (-1 0) are already in *nf* so this step is satisfied. Then, we would have to substitute the two arguments for the related occurrences of bound variables. The first operand is duplicated; the second is dropped. The final step, after noting that the operand is again in *whnf*, would be to substitute it in the body of the term producing the result (-1 (-2 1 0) 0). As one can see, several nested "inspections" are implied by such an informal procedure. The introduction of closures and environments makes mechanization direct and straightforward. The *SECD* reducer discussed in the following section makes this aspect more concrete.

0.	((-2 1 1) (-2 1 0) (-1 0))	*e*
1.	(-2 1 0)	*whnf*
2.	(-1 0)	*whnf*
••3.	((-2 1 0) (-2 1 0))	
4.	(-2 1 0)	*whnf*
•5.	(-1 (-2 1 0) 0)	*whnf*

Table 3.2 Eager reduction of ((-2 1 1)(-2 1 0)(-1 0)).

It is remarkable that, at an abstract level, the *eval-apply* interpreter encompasses the basic function-calling mechanisms of both imperative languages and traditional applicative languages like *Lisp* or *Scheme*. The runtime structures of procedure invocation for an *ALGOL*-family language (i.e., a call stack and an environment display) correspond directly to such interpreter's minimal implementation requirements. Naturally, implementations of such languages are additionally burdened by the complications introduced by box-variables. Call-by-name, another feature of *ALGOL*, corresponds to normal-order evaluation and it is not directly present in the basic *eval-apply* strategy.

―――――――――

[17] Such a strategy in akin to implementations of "call by value" that abound in most current languages.

Essence of Graph Reduction

Now we elaborate on the previous comment about what constitutes the essence of graph reduction. Reduction under *EV-AP* in the linear *BTF* notation (cf. Table 3.1) appears to duplicate expressions. But this need not be the case. The linear notation is used simply because it helps in showing operational detail. In practice, *eval-apply* operates on REP-style graphs. At each step, the environment consists of a pair of pointers to the graph representation of the term. For instance, the final *whnf*, which in graph parlance is written

$$(clos \ \#3=(lam \ \#1 \ (app \ \#2 \ \#1\#)) \ ((\#2\# \ clos \ (lam \ \#2\# \ \#3\#) \ nil))),$$

actually shares the subgraph *(lam #1 (app #2 #1#))* or (-1 1 0) between the term and environment portions of the outer closure (Figure 3.1). Node #3# is common between the two closures. Therefore, given suitable representations, even the simplest of reduction algorithms actually constitute graph reduction. For the interested reader, the complete sequence in linear graph form appears in Table B.8.

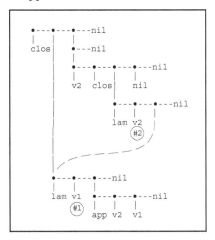

Figure 3.1 Result graph of EV-AP reduction with pointer variables v1, v2 for nodes #1 and #2.

Style of Concrete Definitions

The format of *Lisp* definitions employed in all the abstract reducers is shown next (Figure 3.2). With the exception of a few and local surgical updates, all the reducers are given in what is essentially a pure, albeit a bit old-fashioned, functional style[18]. The constructors and accessors used were exemplified in section 2.2. A listing of a concrete but minimal definition for each reducer can be found in Appendix A.

```
(defun eval (term env)

  (if (is-var term)
    (val-of (look-up term env))

    (if (is-app term)
      (apply (eval (fun-of term) env) (eval (arg-of term) env))

      (if (is-lam term)
        (mk-clos term env)))))

(defun apply (rator rand)

  (let ((fun (term-of rator))
        (env (env-of rator)))
    (eval (form-of fun) (mk-bind (bnd-of fun) rand env))))

(defun look-up (var env)
  (if (eq var (var-of env))
    env
    (look-up var (rest-of env))))
```

Figure 3.2 McCarthy's Eval-apply Interpreter as Lambda Calculus Reducer.

3.3 LANDIN'S SECD MACHINE

SECD stands for *Stack, Environment, Control* and *Dump*. An algorithm which defines state transitions between these four elements is suited for weak reduction of a Lambda Calculus *wfe*. This machine was described in the seminal paper of Peter Landin "The Mechanical Evaluation of Expressions" [Landin 64]. Landin is a pioneer of applicative languages and of their relation to the Lambda Calculus. He also used the calculus to explain the constructs of the *ALGOL* family of languages [Landin 65].

The four elements of the *SECD* machine play the roles implied by their names. In particular, the control is where the expression under reduction is placed. All four are managed as LIFO lists. If we employ a de Bruijn representation for variables then the lookup procedure is an indexing operation into the environment list.

We proceed now with the details. An abstract machine description of the *SECD* algorithm is shown in Table 3.3; a concrete reducer appears in Figure A.2.

---

[18] This style is identical to the one that Aiello and Prini utilized in the presentation of their Lambda Calculus reducer [Aiello 81]. We have to commend them for offering a concrete style that is expressive and readable—even while having many parentheses.

Stack	Env	Cntl	Dump	Stack	Env	Cntl	Dump
$\lambda a\{\rho\}{:}\varnothing$	{}	{}	{}	—	—	$STOP$	—
s	ρ	$\lambda\,a\cdot c$	d	$\lambda a\{\rho\}{:}s$	ρ	c	d
s	ρ	$ab\cdot c$	d	s	ρ	$b\cdot a\cdot @\cdot c$	d
s	$u_0\bullet\ldots\bullet u_i\bullet\ldots\bullet\rho$	$i\cdot c$	d	$u_i{:}s$	$u_0\bullet\ldots\bullet u_i\bullet\ldots\bullet\rho$	c	d
$\lambda a\{\rho'\}{:}C{:}s$	ρ	$@\cdot c$	d	$\varnothing$	$C\bullet\rho'$	$a\cdot\varnothing$	$\langle\ s,\rho,c,d\rangle$
$\lambda a\{\rho\}{:}\varnothing$	{}	{}	$\langle s,\rho,c,d\rangle$	$\lambda a\{\rho\}{:}s$	ρ	c	d

Table 3.3 Abstract machine for the eager SECD reducer.

The problem expression is the template which controls the process. If a is the term to be reduced, the machine is started with the quadruplet $\langle\varnothing,\varnothing,a,\varnothing\rangle$. As each subterm is processed it is removed from the control stack. Variables are looked-up and abstractions are immediately pushed onto the stack as closures. Applications are broken apart to their operator and operand subterm and each is separately reduced to *whnf*. Once they are reduced they are applied to each other—exactly like in the case of *eval-apply*. This is signaled by a special marker that is pushed onto the control stack when an application node is processed. A single complication arises; when the topmost closure on the stack is applied to the one

step	stack	env	cntl	dump
S0	nil	nil	(((-2 1 1) (-2 1 0) (-1 0)))	nil
S1	nil	nil	((-1 0) ((-2 1 1) (-2 1 0)) @)	nil
S2	([(-1 0) { }])	nil	(((-2 1 1) (-2 1 0)) @)	nil
S3	([(-1 0) { }])	nil	((-2 1 0) (-2 1 1) @ @)	nil
S4	([(-2 1 0) { }][(-1 0) { }])	nil	((-2 1 1) @ @)	nil
S5	([(-2 1 1) { }][(-2 1 0) { }][(-1 0) {}])	nil	(@ @)	nil
•S6	nil	([(-2 1 0) { }])	((-1 1 1))	(:s ([(-1 0) { }]) :e nil :c (@) :d nil)
S7	([(-1 1 1) {[(-2 1 0) { }]}])	([(-2 1 0) { }])	nil	(:s ([(-1 0) { }]) :e nil :c (@) :d nil)
S8	([(-1 1 1) {[(-2 1 0) { }]}][(-1 0) {}])	nil	(@)	nil
•S9	nil	([(-1 0) { }][(-2 1 0) { }])	((1 1))	(:s nil :e nil :c nil :d nil)
S10	nil	([(-1 0) { }][(-2 1 0) { }])	(1 1 @)	(:s nil :e nil :c nil :d nil)
S11	([(-2 1 0) { }])	([(-1 0) { }][(-2 1 0) { }])	(1 @)	(:s nil :e nil :c nil :d nil)
S12	([(-2 1 0) { }][(-2 1 0) { }])	([(-1 0) { }][(-2 1 0) { }])	(@)	(:s nil :e nil :c nil :d nil)
•S13	nil	([(-2 1 0) { }])	((-1 1 0))	(:s nil :e ([(-1 0) { }] [(-2 1 0) { }]) :c nil :d (:s nil :e nil :c nil :d nil))
S14	([(-1 1 0) {[(-2 1 0) { }]}])	([(-2 1 0) { }])	nil	(:s nil :e ([(-1 0) { }] [(-2 1 0) { }]) :c nil :d (:s nil :e nil :c nil :d nil))
S15	([(-1 1 0) {[(-2 1 0) { }]}])	([(-1 0) { }][(-2 1 0) { }])	nil	(:s nil :e nil :c nil :d nil)
S16	([(-1 1 0) {[(-2 1 0) { }]}])	nil	nil	nil

Table 3.4 SECD-V reduction of ((-2 1 1)(-2 1 0)(-1 0)) .

behind itself a new cycle must begin. The state of the machine must be remembered by "dumping" it to the appropriately named auxiliary structure. This step is required because any pending elements of the stack must be restored and processed in the applicable environment when the result of the reduction becomes available.

Table 3.4 shows a trace of this algorithm on the example expression that was utilized in the previous section. It looks more complicated but the reader is reminded that this is an algorithm that does not rely on the implicit recursive call stack of *eval-apply*.

This is the simplest version of the *SECD* machine. Its admirable attribute is its sequencing which ensures that all the elements are available when and where they are needed. A simple automaton with four stacks is all that is necessary. In a more applied vein, an eager applicative language can be implemented using the concepts of this machine. Peter Henderson's *Lispkit* [Henderson 80], [Henderson 82] defines a *compiling* procedure which can parse an applicative expression and produce an equivalent sequence of control instructions. When these instructions are allowed to drive an empty machine they produce more efficiently the same result as the *eval-apply* interpretation of the applicative expression.

Variations of the *SECD* machine make it possible to evaluate subterms only when they are referred to by a variable or when they are the arguments of a strict primitive. We will not discuss the details of such versions here because, even though historically influential, they are actually quite cumbersome. The mechanisms of Wadsworth/Aiello–Prini's *RTLF* and the equivalent weak version of *HOR* are much more suited to weak normal-order reduction.

As a Lambda Calculus abstract reducer, the *SECD* machine appears to always expend more effort than *eval-apply* even though it is a mechanization of its tactic. One should keep in mind that what we are really doing is "simulating" a machine capable of function evaluation on top of an already capable framework. The *SECD* machine is by itself a device which, given a simple underlying automaton, provides reduction and evaluation mechanisms for functions. This cannot be claimed for *eval-apply* whose definition, as presented in the previous section, relies on an underlying language that consists of applicative recursive function evaluation.

Another class of reducers of the naturally weak kind is the Categorical Abstract Machine (*CAM*) [Cousineau 85]. It is based on the syntactic equivalence between the Lambda Calculus and a form of Combinatory logic founded on *Cartesian closed categories*. Like *eval-apply* and the *SECD* machine, *CAM* in its native form follows an applicative order strategy.

We are now ready to continue with reduction that advances beyond weak head normal form. In terms of the classification of Table 1.6 we are now crossing the division in the middle. Reduction to head normal form presents practically all the difficulties of strong reduction so it is not discussed separately. For a *hnf*-specific reducer see section 4.1.

3.4 WADSWORTH REDUCTION

... The advantage of the literal substitution model is that [the] substituted expression is immediately available at places where it is needed. The disadvantage, of course, is that some copying must be done. The suspended substitution model has exactly the opposite properties; [a] substituted expression is not made available immediately, but copying is done in a delayed fashion. ... [Kathail 90].

Orthogonal to the idea of delaying substitution as long as possible, is Wadsworth's graph-based method of literal argument substitution. This method has been so influential that when graph reduction for the Lambda Calculus is discussed, Wadsworth's technique is commonly implied. Christopher Wadsworth presented his system in the final chapter of his Ph.D. thesis [Wadsworth 71]. It has since been the standard by which new proposals are judged. His work still offers, both in terms of representation and overall strategy, a mix of ideas that are important for efficient Lambda Calculus reduction.

In Wadsworth's technique, β-reduction is effected solely by manipulating pointers. Each substitution is fully effected in a graph which represents the expression under reduction. A redex is contracted by overwriting the root node of the redex by a pointer which points to the abstraction's body. The abstraction node itself is overwritten by an indirection arc pointing to the argument (Figure 3.3). Occurrences of the abstraction's binding variable in the body of the abstraction see the effect automatically by following the indirection arcs of their binders.

RTNF/RTLF strategy

The preceding description of how redices are contracted covers only a small portion of the necessary steps of a complete reduction algorithm. We have not touched upon yet the question of how to organize the traversal of the graph, which redex to contract, and whether the simple "rewire the pointers" operation is sufficient for β-reduction.

Wadsworth gave the basis of the recursive strategy he had employed as well as various hints about its implementation. We only sketch the core idea here since a delayed version of this strategy which is closer to our overall approach is described in detail in the next section.

The function of a top-level application is reduced enough to expose a topmost abstraction, if one exists. In this case the strategy is applied recursively to the body of the new abstraction after substituting the original application's argument for the topmost abstraction's bound variable. In case the reduction of the application's function does not reveal a topmost abstraction, the topmost application remains as part of the overall normal form and reduction proceeds with its argument. Reducing only enough to reveal a topmost abstraction—i.e., observing a weak strategy—is isomorphic to the overall strong reduction strategy except for the case of an abstraction which is returned as it stands. In contrast, under the overall strong strategy, an unapplied abstraction structure is duplicated, thereby introducing a new abstraction whose body is the recursively strongly reduced body of the unapplied abstraction. This is the essence of the so called *RTNF/RTLF* strategy—reducing to the <u>n</u>ormal <u>f</u>orm or the <u>l</u>ambda <u>f</u>orm of the expression.

Copying of Operator Graphs

The mechanisms of β-contraction effected through pointer switching, and the above-sketched *RTNF/RTLF* strategy are still insufficient for correct reduction. An additional requirement is present when an abstraction is shared between multiple application nodes in operator position. This complication is illustrated by utilizing the same example expression, $(\lambda x.bx \ (\lambda z.za)) \ (\lambda x.bx \ (\lambda z.za))$, as the one in Wadsworth's fourth chapter.

If one follows verbatim the rule for beta contraction on the graph of Figure 3.4 (*A*) then the graph (*B*) is produced. This graph is incorrect because the operator of the top redex is pointed to by two distinct application nodes. In such a failed attempt one is left with a cyclic graph which does not represent an expression to which the original expression is reducible. The resulting graph is not even admissible as a *wfe*. An alternative way of recognizing the inappropriateness of the situation is by observing that the free occurrences of the bound variable in the operator will need to be substituted for two *different* operand expressions.

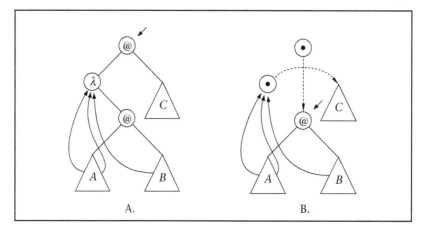

Figure 3.3 Wadsworth's beta contraction via pointer manipulation.

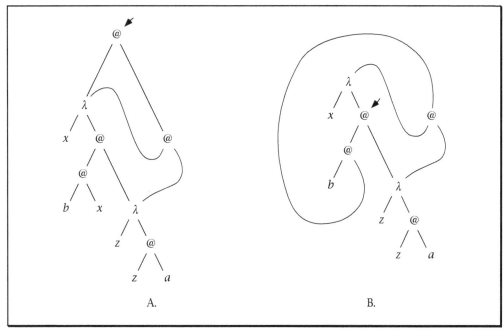

Figure 3.4 Non-admissible graph for (λx.bx (λz.za))(λx.bx (λz.za))(λz.za)).

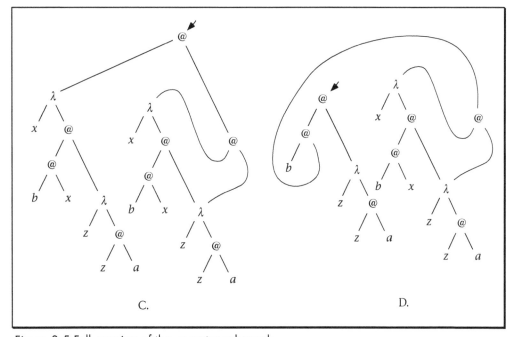

Figure 3.5 Full copying of the operator subgraph.

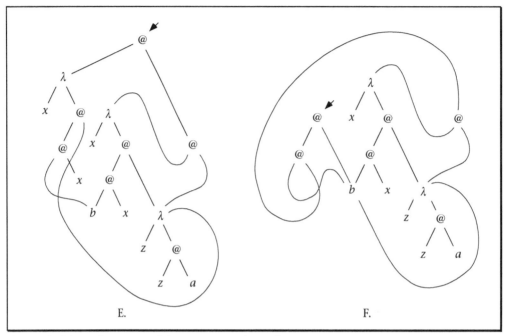

Figure 3.6 Partial copying of the operator subgraph.

A non-admissible graph can be turned to an admissible one by copying the operator subgraph, as in the graph of Figure 3.5 (*C*), thus ensuring that the operator node of the top redex is pointed only by the top redex itself. After one contraction, the resulting graph is like (*D*) in the last figure.

It is not the case though that all nodes of the shared operator need to be copied. One can avoid copying any nodes (and their descendants) of the operator in which the bound variable of the abstraction does not occur free (see graphs of Figure 3.6 (*E*) and (*F*)). Such nodes belong to what are termed *maximally free expressions* (*mfe's*) which can remain shared if judicious copying is done. It is not difficult to see that such selective copying may become quite expensive since it requires that an up-to-date account of all *mfe's* is maintained throughout the process.

In all, Wadsworth's strategy was a marked improvement of the state of Lambda Calculus reduction and it is justifiably regarded as a major achievement. The unavoidable copying leads to some loss of sharing and results in extra reductions that could be avoided with a more elaborate strategy for choosing redices. This aspect of the *RTNF/RTLF* strategy is revisited in section 6.2 when the augmentation of the delayed substitution algorithms with sharing is explored.

3.5 Aiello–Prini Environment Graph Reduction

The next reducer in our presentation combines the techniques and shares the advantages of the *eval-apply* and Wadsworth families. It is our belief that before the introduction of Head Order Reduction this class of normal order graph reduction stood unmatched in terms of meeting the goals set forth in the introductory chapter of this thesis. When augmented with sharing, this strategy may still be the best for many classes of lambda expressions.

There is a pronounced similarity to Wadsworth's graph reduction algorithm. Wadsworth ended his thesis with the words:

> … Judged in the terms of the introduction to this [4th] chapter, graph reduction extends the range of simulated substitution mechanisms. But the technique has enough novelty to merit a search for equivalent interpreters/fixed program machines [Wadsworth 71].

Aiello and Prini responded to Wadsworth's first quest by emphasizing delayed substitutions and organizing the recursive functions *RTNF/RTLF* around the concept of the environment. Our recasting of this strategy with the HOR machinery of de Bruijn variables and *UVC*s as the reducer *AP-HOR* is a further development (section 4.6) along this direction.

Luigia Aiello and Gianfranco Prini's reducer is developed in a sequence of increasingly sophisticated versions starting with a naive, literal substitution algorithm [Aiello 81]. They show (very graphically, using a metaphor of surgeons finding and removing bladder stones!) how delaying substitutions can lead to an efficient reducer. The top-level of their final exposition takes the form of two mutually recursive functions directly analogous to Wadsworth's *RTNF/RTLF*.

AP Reduction Narrative

Aiello–Prini's reducer follows Wadsworth's strategy which is defined via the mutually recursive functions *rtnf* and *rtlf*. The former returns *nf*'s and the latter operational *whnf*s. They are isomorphic except for the treatment of abstractions.

The core idea of this strategy is to effect enough reductions on an operator expression until its top-level abstraction is revealed, if one is present, and to then effect a reduction by combining the abstraction with the first applicable operand. This means that *rtlf* alone is a *whnf* strategy. Delayed substitutions are effected via the usual suspension/environment constructions.

We proceed now with a detailed case-by-case description of the algorithm.

Procedure rtnf (term env)

If the term is a **variable**, we look it up on the current environment. If it dereferences to a suspension then strong normalization continues with the term and the environment portion of the suspension. Otherwise it must dereference to a relatively free variable introduced when an unapplied abstraction was traversed. In this case the dereferenced variable is returned.

In case the term is an **application**, we first attempt to weakly reduce the operator of the application via *rtlf*. If we succeed we simulate a reduction by continuing with strong normalization of the body of the *whnf* (returned lazily as a suspension-environment pair) and an environment which is produced by adding a binding which consists of the abstractor of the *whnf* with the suspension made from the operand of the application together with the current environment. If *rtlf* does not return a suspension, then the operator is already in *nf* and strong normalization continues with the operand.

In case the term is an **abstraction**, the sequencing of *rtnf* guarantees that the abstraction is unapplied. A fresh variable is introduced and we make a true suspension by pairing this variable with the current environment. Strong normalization continues with the term of the abstraction and an environment which is the current one augmented with the binding we just made.

Procedure rtlf (term env)

This procedure is identical to *rtnf* with two exceptions:

First, in the case of an **abstraction**, the abstraction is returned immediately as a suspension-environment pair since it is already a *whnf*. Second, in all instances but one, when *rtnf* specifies continuation of strong normalization, *rtlf* specifies continuation of weak normalization. The one exception is the case of the normalization of an operand of an application whose operator is already in *nf*. This normalization must be strong in *rtlf* also. In other words, *rtlf* has calls to itself in the places where *rtnf* has calls to itself except for the case just mentioned.

A Compact AP Machine

Since the *rtnf* and *rtlf* functions are nearly identical to each other, they can be cast as a single recursive function (Figure 3.7). The two modes are incorporated in a compact, folded form via a third flag argument which determines the "strength" of the normalization. We call this flag a *force* flag. When true, it allows strong normalization of its term argument. When false, it limits reductions to a lazy weak head normal form. The conciseness of this strongly normalizing algorithm is indeed remarkable.

```
(defun rtnlf (term env f)

  (if (is-var term)
    (let ((susp (val-of (look-up term env))))
      (if (is-susp susp)
        (rtnlf (term-of susp) (env-of susp) f) susp))

    (if (is-app term)
      (let ((susp (rtnlf (fun-of term) env nil)))
        (if (is-susp susp)
          (rtnlf (frm-of (term-of susp))
                (mk-bind (bnd-of (term-of susp))
                      (mk-susp (arg-of term) env) (env-of susp)) f)
          (mk-app susp (rtnlf (arg-of term) env t))))

      (if f
        (let ((v (mk-var '\_)))
          (mk-lam v (rtnlf (frm-of term) (mk-bind (bnd-of term) v env) t)))
        (mk-susp term env)))))
```

Figure 3.7 Compact RTNF/RTLF normal order reducer with a "force" flag.

AP Reduction Example

Now, let us follow how the plain *AP* reducer works on Wadsworth's example expression *wads45* = λab.(λc.c (c a)) ((λde.e d) b) which in *BTF* is written as (-2 (-1 0 (0 2)) ((-2 0 1) 0)). Each step is labeled with a numbered instance of *N*, *L* or *R*. The first two denote invocations of *rtnf* and *rtlf* respectively; the last one signifies a corresponding return. The complete trace of *wads45* via *AP* appears in Table B.5. Below, we highlight some of the most characteristic portions.

First, when outer unapplied abstractions are traversed, we push freshly-created unbound marker variables onto the environment:

```
N0:   (-2 (-1 0 (0 2)) ((-2 0 1) 0))        ()
N1:   (-1 (-1 0 (0 2)) ((-2 0 1) 0))        (g1)
N2:   ((-1 0 (0 2)) ((-2 0 1) 0))           (g2 g1)
```

If the operator of an application is an abstraction, the abstraction is returned immediately as a closure:

```
N2:   ((-1 0 (0 2)) ((-2 0 1) 0))           (g2 g1)
L3:   (-1 0 (0 2))                          (g2 g1)
R3:   ([ (-1 0 (0 2)) ({ g2 g1 }) ])
```

The returned closure participates in a beta reduction. This has the effect of pushing a new closure formed from the application's argument and the current environment onto the environment. Reduction continues with the body of the abstraction:

```
N2:  ((-1 0 (0 2)) ((-2 0 1) 0))        (g2 g1)
L3:  (-1 0 (0 2))                       (g2 g1)
R3:  ([ (-1 0 (0 2)) ({ g2 g1 }) ])
•N4: (0 (0 2))                          (([ ((-2 0 1) 0) ({ g2 g1 }) ]) g2 g1)
```

A variable is looked up on the current environment. If the returned term is another variable, the process repeats itself. All the while, the "strength" of the reduction strategy, *nf* or *whnf*, (N or L in the traces) is observed.

```
L17:  0      (([ 2 ({ ([ ((-2 0 1) 0) ({ g2 g1 }) ]) g2 g1 }) ]) ([ 0 ({ g2 g1 }) ]) g2 g1)
L18:  2      (([ ((-2 0 1) 0) ({ g2 g1 }) ]) g2 g1)
R18:  g1
R17:  g1
```

Finally, when all the recursive calls to the strategy have been completed, the unbound lambda marker variables are looked up on the environment that was in effect at the time of the call and by doing so are related to the enclosing topmost abstractions. The named global variables result from the *r2b* translation to *BTF* (cf. section 2.4).

```
R2:  (g1 g2 g2)
R1:  (-1 g1 0 0)
R0:  (-2 1 0 0)
```

An Aiello–Prini reducer, including sharing, and modified for compactness is shown in Figure A.4.

3.6 BERKLING'S HEAD ORDER REDUCTION

The difficulty of performing reductions in the presence of free variables is well recognized in the computer science community [PeytonJones 87], [Field 88]. Many researchers have been willing to avoid strong normalization because of this difficulty. But, as hinted by de Bruijn and advocated by Berkling over the years, there exists an elegant solution to this problem. A single register needs to be added to the runtime machinery of reduction. This register holds at all times a count of the number of times the composite substitutions, represented by the state of the environment, has been carried over unapplied abstractions. With such an element in place, determining the correct de Bruijn index of unbound variable occurrences becomes a unit time operation which can be performed on-demand as an expression is traversed.

In order to give a presentation of Head Order Reduction which is not only instructive but also true to its intellectual origins we first show the graphical aids that Berkling has suggested and extensively used.

Slanted line segments connected together in a zig-zag line symbolize nested abstractions interspersed with nested applications. Under this convention the focus is on the left spine of an expression. Right-pointing corners are formed when abstractions are followed by applications. The opposite is true for the left-pointing corners (Figure 3.8).

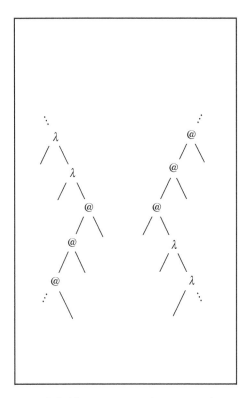

Figure 3.8 Abstraction-application and application-abstraction corners of a left spine.

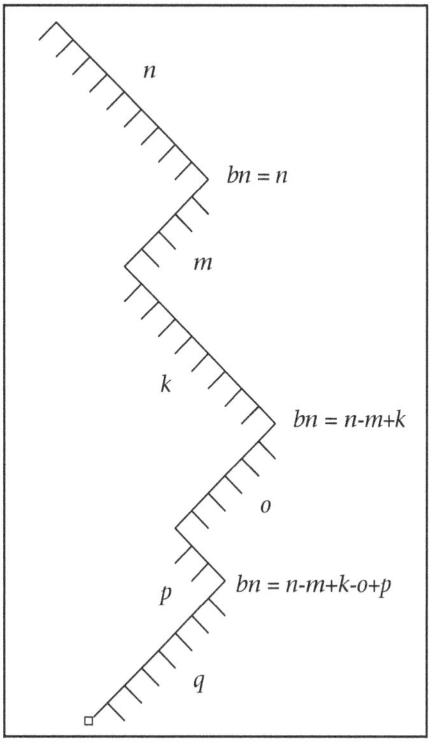

Figure 3.9 Left Spine tagged with a count of binding levels.

Each sequence of applications can be tagged with a count of the uninstantiated bindings along the left spine (Figure 3.9). For each unapplied abstraction the count of yet-to-be-bound variables (binding number, bn) is incremented by one. This has the effect that when an unapplied abstraction is referred to, the correct de Bruijn index that would point to this abstraction can be calculated easily. It is the difference of the total number of abstractions minus the number of operands encountered along the path. This is a vital observation because it permits us to have at all times an immediate mechanism for referring via a de Bruijn index to unapplied abstractions. It also allows us to arrange for "navigating" an expression in an equivalence-preserving manner by pushing adjusting dummy variables for the missing arguments. This mechanism is further illustrated in section 4.5.

HOR in-the-large

The symbolic depiction of Figure 3.10 is helpful in visualizing the three rewrite rules of Head Order Reduction. These rules operate on whole portions of a lambda expression and hence are termed rules of HOR "in-the-large". The overall objective of the application of these rules is to transform an expression to a lazy head normal form.

Figure 3.10 (A) symbolically shows η-extension in-the-large. If there are insufficient arguments to completely instantiate a sequence of abstractions, a bulk η-extension is performed to generate the proper number of artificial ones. The equal number of abstractions and the respective bound variables is represented by the small isosceles triangle.

A symbolic depiction of β-distribution in-the-large appears in Figure 3.10 (B). A sequence of abstractions together with an equal number of arguments comprise a set of delayed bindings which may be recorded for later use. The distribution results in the removal of a left-pointing corner from the spine of the expression. Equivalence is preserved by prepending the removed corner to all the subterms emanating from the spine that are situated below the location of the removed corner.

Finally, Figure 3.10 (C) symbolically shows identity-reduction in-the-large. The head variable selects one of the arguments—a composite expression consisting of an unreduced portion of the problem graph prepended with a recursively formed environment expression—and the rest of the structure disappears. Identity-reduction in-the-large is the formal counterpart of an environment lookup. The integer binding indices make this duality entirely transparent.

A formal definition of these rules, termed η_{ext}, β_{dis} and β_h profits from a casting in *BTF* notation because variable name scoping rules can be avoided. A more traditional definition is given by Zhang and Berkling in [Zhang 89].

Eta Extension in-the-large: η_{ext}

$$((-m\ e_0)\ e_1\ \ldots\ e_n) \rightarrow (-m+n\ (-m\ e_0)\ e_1\ \ldots\ e_n\ m-n-1\ \ldots\ 0),\quad m>n \quad (3.2)$$

Beta Distribution in-the-large: β_{dis}

$$((-m\ e_0\ e_1\ \ldots\ e_k)\ E_1\ \ldots\ E_n) \rightarrow e_0'\ e_1'\ \ldots\ e_k'\ E_{m+1}\ \ldots\ E_n$$
$$\text{where}\ e_i' = ((-m\ e_i)\ E_1\ \ldots\ E_m),\quad 0 \le i \le k,\ m \le n \quad (3.3)$$

Identity Reduction in-the-large: β_h

$$((-m\ i)\ e_{m-1}\ \ldots\ e_0) \rightarrow e_i,\quad i < m \quad (3.4)$$

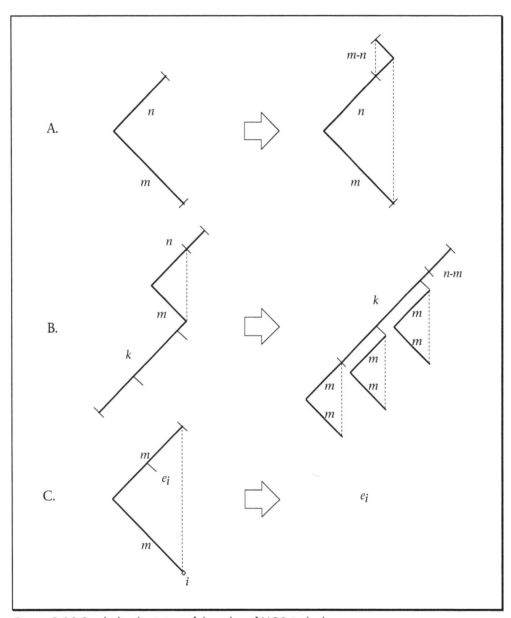

Figure 3.10 Symbolic depiction of the rules of HOR in-the-large.

Example of in-the-large rules

First, we illustrate briefly one application of each of the in-the-large rules.

A composite abstraction which expects five arguments but which is supplied only with three, is transformed via (3.2) as follows:

$$((-5 \ e_0) \ e_1 \ e_2 \ e_3) \to (-2 \ (-5 \ e_0) \ e_1 \ e_2 \ e_3 \ 1 \ 0)$$

Via rule (3.3) a composite abstraction which expects two arguments but which is supplied with three, absorbs the two arguments and distributes them over all subterms:

$$((-2 \ e_0 \ e_1 \ e_2 \ e_3 \ e_4) \ E_1 \ E_2 \ E_3) \to e_0' \ e_1' \ e_2' \ e_3' \ e_4' \ E_3$$
$$where \quad e_0' = ((-2 \ e_0) \ E_1 \ E_2), \ \dots, \ e_4' = ((-2 \ e_4) \ E_1 \ E_2)$$

Finally, an abstraction with a single variable as its body acts as a selector and, according to rule (3.4) directly retrieves the corresponding argument[19]:

$$((-4 \ 1) \ e_3 \ e_2 \ e_1 \ e_0) \to e_1$$

Next, we show an example of how the in-the-large rules are normalizing even while postponing β-reduction as long as possible. The reader should be reminded that literal application of these rules is not, and was not meant to be, an efficient reduction scheme.

Table 3.5 is a step-by-step application of the in-the-large rules to the expression ((-1 0 0) (-2 1 0)). Each occurrence of a rule's left side is underlined. The application of the rules is straightforward. The only β-reductions effected are the identity ones which happen during the in-the-large selection of an argument via a head variable. But the reader is asked to observe the following detail.

In step (7) the trailing free variable with a de Bruijn index of zero (shown in bold type) had to be incremented by one when the η-extension was performed. This variable was introduced via an η-extension that occurred in step (3).

It is indeed true that the only variables that need "correction" under Berkling's in-the-large rules are the ones referring to dummy bindings introduced via an η-extension. The amount of necessary correction is determined by the aggregate size of all intervening η-extensions; i.e., the number of new abstractions and of the matching variables that were introduced. In this example both η-extensions are of unit size since they consist of a single new abstraction-variable pair. We illuminate this point further in section 4.2.

---

[19] An environment lookup will succeed if and only if the head variable is "within range". This is guaranteed when the in-the-large rules are applied to a closed, artificially or not, expression.

0.	((-1 0 0) (-2 1 0))	
1.	(((-1 0) (-2 1 0)) ((-1 0) (-2 1 0)))	β_{dis}
2.	((-2 1 0) ((-1 0) (-2 1 0)))	β_h
3.	(-1 (-2 1 0) ((-1 0) (-2 1 0)) 0)	η_{ext}
4.	(-1 ((-2 1) ((-1 0) (-2 1 0)) 0) ((-2 0) ((-1 0) (-2 1 0)) 0))	β_{dis}
5.	(-1 ((-1 0) (-2 1 0)) ((-2 0) ((-1 0) (-2 1 0)) 0))	β_h
6.	(-1 (-2 1 0) ((-2 0) ((-1 0) (-2 1 0)) 0))	β_h
7.	(-1 (-1 (-2 1 0) ((-2 0) ((-1 0) (-2 1 0)) 1) 0))	η_{ext}
8.	(-1 (-1 ((-2 1) ((-2 0) ((-1 0) (-2 1 0)) 1) 0) ((-2 0) ((-2 0) ((-1 0) (-2 1 0)) 1) 0)))	β_{dis}
9.	(-1 (-1 ((-2 0) ((-1 0) (-2 1 0)) 1) ((-2 0) ((-2 0) ((-1 0) (-2 1 0)) 1) 0)))	β_h
10.	(-1 (-1 1 ((-2 0) ((-2 0) ((-1 0) (-2 1 0)) 1) 0)))	β_h
11.	(-1 (-1 1 0)) = (-2 1 0)	β_h

Table 3.5 Example of transforming an expression via the in-the-large rules.

Head Normal Forms

Recall from equation (2.5) in section 2.1 that a *hnf* is of the form

$$M = \lambda x_1 ... x_m . x_i \, e_1 ... e_n \quad \text{where} \quad 1 \le i \le m.$$

In *BTF* notation, the same expression is written as

$$M = (- m \;\; m - i \;\; e_1 \;\; e_2 \;_{...} \; e_n), \;\; 1 \le i \le m.$$

Figure 3.11 shows that the spine of such an expression consists of a single, right-pointing corner with the head variable at the bottom of the spine.

Meaning of Suspensions and Environments

The transformation rules of HOR prompt us now to rethink the meaning of the operational constructs of suspensions and environments in terms of the Lambda Calculus. In general an environment is a mapping from a vector $\vec{x}$ of free variables to a vector $\vec{e}$ of "values". A suspension is a pairing of an open lambda term together with an environment which instantiates the term's relatively free variables (3.5).

$$\rho = \langle \vec{x}, \vec{e} \rangle$$
$$[a \;\; \rho] = (\lambda \vec{x}.a)\vec{e}$$

(3.5)

With this insight, the meaning of a suspension $[a \;\; \rho]$ is that of the environment ρ prepended to the term a. Figure 3.12 shows the construction in two dimensions. In terms of the concept of *contexts* an environment and a suspension can be defined as in (3.6).

$$C_\rho[\ ] = (\lambda \vec{x}.[\ ])\vec{e}$$
$$[a\ \rho] = C_\rho[a]$$

(3.6)

Now, it is instructive to visualize the overall reduction of a lambda expression as the successive straightening of the zig-zag lines representing its left spine and, recursively, those of the delayed subterms emanating from the straightened left spine.

As a lambda expression is traversed along its left spine in a downward direction; i.e., from the root to the head variable, the abstraction-argument pairs are removed and we note the proper bindings for subsequent variables resulting from these reductions. The bookkeeping takes the form of nested abstraction-application triangles as shown in Figure 3.13 (*B*). When the head variable is finally encountered, a *lazy* or *operational hnf* is reached. It consists of a sequence of topmost abstractions followed by a sequence of suspensions and culminates in a head variable.

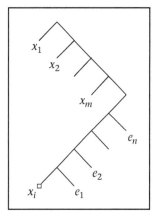

Figure 3.11 Picture of a Head Normal Form.

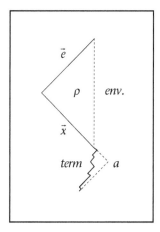

Figure 3.12 Meaning of a suspension as a Lambda Calculus *wfe*.

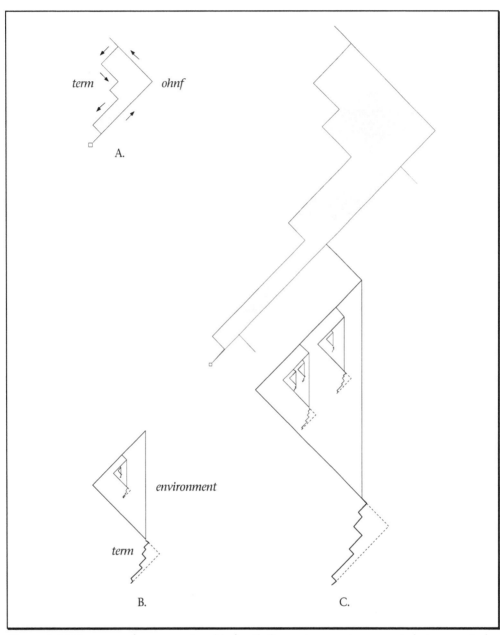

Figure 3.13 Structure of an operational hnf with the environment represented via recursively "pushed" abstraction-application triangles.

Weak Head-Order Reduction

Before the strongly normalizing algorithm which is the focus of Head Order Reduction is considered, we present an intermediate reducer for obtaining a weak head normal form. This detour is helpful because it gives the opportunity to introduce a simplified abstract machine that is a subset of the full reducer. The strong normal order reducer is obtained by solving the problem of reductions in the scope of relatively free variables and by arranging for recursive calls on the basic strategy.

Term	Stack	Env	Term	Stack	Env
λa	$\{\}$	ρ	$\lambda\, a\{\rho\}$		
λa	$a'\{\rho'\}\!:\!s$	ρ	a	s	$a'\{\rho'\}\bullet\rho$
ab	s	ρ	a	$b\{\rho\}\!:\!s$	ρ
$i+1$	s	$u\bullet\rho$	i	s	ρ
0	s	$a\{\rho'\}\bullet\rho$	a	s	ρ'

Table 3.6 Weak Head Order Reduction Abstract Machine.

The abstract machine for this weak HOR[20] reducer (*W-HOR*) has five rules (Table 3.6). It is completely deterministic because all the cases are mutually exclusive. To reduce a term a the machine is started with the triplet $\langle a,\{\},\{\}\rangle$. When the stack is empty the machine stops returning the closure formed by pairing the final term with the last environment in effect. If the last two cases which specify the lookup procedure are implemented by an indexing operation into the environment structure then the result has a DAG structure.

W-HOR appears to us to be the simplest weak reducer with a normal-order strategy. As such it must have been described before. This reducer is indeed identical to the "*λ-RED*" of Oberhauser [Oberhauser 87] except for the fact that it is being presented here in a de Bruijn notation[21]. According to the commonly accepted terminology, and in contrast with the claim made in the last reference, this is a *whnf* reducer. The immediate benefit of de Bruijn representation for variables is that the environment lookup of the last two rules is, in the abstract and to a certain degree in a reasonable implementation, a direct indexing operation.

---

[20] Strictly speaking, it is not correct to call this reducer "Head-Order". Traversal proceeds towards the head of an expression but only as long as the top-level applications are not exhausted. Nevertheless, the name has stuck with us since this reducer is a proper subset of strong HOR.

[21] We believe *W-HOR* is also identical to *Krivine's Machine* but to date we have not located a proper description.

step	term	stack	env
D0	((-2 1 1) (-2 1 0) (-1 0))	nil	nil
D1	((-2 1 1) (-2 1 0))	(([(-1 0) ({ })]))	nil
D2	(-2 1 1)	(([(-2 1 0) ({ })]) ([(-1 0) ({ })]))	nil
•D3	(-1 1 1)	(([(-1 0) ({ })]))	(([(-2 1 0) ({ })]))
•D4	(1 1)	nil	(([(-1 0) ({ })]) ([(-2 1 0) ({ })]))
D5	1	(([(-2 1 0) ({ })]))	(([(-1 0) ({ })]) ([(-2 1 0) ({ })]))
D6	(-2 1 0)	(([(-2 1 0) ({ })]))	nil
•D7	(-1 1 0)	nil	(([(-2 1 0) ({ })]))

Table 3.7 W-HOR reduction of ((-2 1 1)(-2 1 0)(-1 0)).

Table 3.7 shows a trace of weak HOR reduction. The same example expression as in the discussion of *EV-AP* and *SECD* (cf. sections 3.2 and 3.3) is reduced. In comparison with these reducers, the minimality of this algorithm is clear. This example expression is too simple for lack of sharing to reveal any inefficiencies.

Section 6.3 discusses how this reducer can be made much more efficient with a simple modification. This change involved introduces sharing of the same degree as the one provided by Aiello and Prini's weak call-by-need technique (*rtlf* with sharing).

In retrospect, and according to our current knowledge of the relevant literature, this weak head-order reduction algorithm forms the point of departure for Berkling's formulation of strong HOR. A concrete *W-HOR* reducer appears in Figure A.6.

Strong Head-Order Reduction

This is the reduction algorithm which is in the very center of this thesis. It combines the idea of delayed substitution via closures and environments with the mechanism that provides effortless updating of the binding indices which correspond to relatively free variables.

$$
\begin{aligned}
((\lambda a)b)\ \rho\ \phi\ &\rightarrow\ a\ \ b\{\rho\}\cdot\rho\ \ \phi \\
(ab)\ \rho\ \phi\ &\rightarrow\ (a\ \rho\ \phi)(b\ \rho\ \phi) \\
(\lambda a)\ \rho\ \phi\ &\rightarrow\ \lambda(a\ \ \phi-1\cdot\rho\ \ \phi-1) \\
i\ c\cdot\rho\ \phi\ &\rightarrow\ i-1\ \rho\ \phi \\
0\ a\{\rho'\}\cdot\rho\ \phi\ &\rightarrow\ a\ \rho'\ \phi \\
0\ \phi'\cdot\rho\ \phi\ &\rightarrow\ \phi'-\phi
\end{aligned}
$$

(3.7)

Head Order reduction in-the-small is captured via the rewrite rules (3.7). The intended meaning of metavariables *a, b, c, i, ρ* and *φ* is given in Table 3.9. The first three rules cover the cases of a redex, an application and of an unapplied abstraction respectively. The last three rules define the process of looking up the value of the variable denoted by an integer de Bruijn index.

These rules are non-deterministic because they do not specify by themselves an order for evaluation. The first rule must have precedence over the second one. If the second rule is used eagerly the overall reducer is potentially non-terminating[22]. As discussed previously, an inseparable aspect of HOR is that one has to organize the order of the application of these rules to ensure descent of the left spine.

Similar rules for HOR have also been provided by Hilton [Hilton 90b].

HOR Abstract Machine

Table 3.8 presents an abstract machine which achieves the requirement just mentioned. The whole machinery of pure HOR is now in full view. The only new elements that need some explanation are the 'λ' and '@' markers appearing in the type signature of the stack and the direction flag d. The lambda marker provides for the reconstruction of the top-level abstraction context; the join marker delineates the reduction of a delayed subterm. The direction flag controls the overall application of the rules.

Term	Stack	Env	UVC	Dir	Term	Stack	Env	UVC	Dir
ab	s	ρ	ϕ	↓	a	$b\{\rho\}{:}s$	ρ	ϕ	↓
λa	$c{:}s$	ρ	ϕ	↓	a	s	$c \bullet \rho$	ϕ	↓
λa	s	ρ	ϕ	↓	a	$\lambda{:}s$	$\phi - 1 \bullet \rho$	$\phi - 1$	↓
$i + 1$	s	$u \bullet \rho$	ϕ	↓	i	s	ρ	ϕ	↓
0	s	$a\{\rho'\} \bullet \rho$	ϕ	↓	a	s	ρ'	ϕ	↓
0	s	$\phi' \bullet \rho$	ϕ	↓	$\phi' - \phi$	s	—	ϕ	↑
a	$\lambda{:}s$	—	ϕ	↑	λa	s	—	$\phi + 1$	↑
b	$@{:}a{:}s$	—	ϕ	↑	ab	s	—	ϕ	↑
a	$b\{\rho'\}{:}s$	—	ϕ	↑	b	$@{:}a{:}s$	ρ'	ϕ	↓
a	$\{\}$	—	ϕ	↑	a	—	—	—	STOP

Table 3.8 Head Order Reduction Abstract Machine.

The syntactic elements are those of Table 3.9. The metavariable u used in the look-up procedure stands for any syntactically admissible environment element. The definition of a concrete HOR reducer can be found in Figure A.9[23].

Reduction is started with the quintuplet $\langle a, \varnothing, \varnothing, 0, \Downarrow \rangle$. The reader should note that all invocations of the functions *down* and *up* are tail-recursive, a property which reflects the

---

[22] A concrete reducer derived by following naively the rules (3.7) is listed in Figure A.3. It is both incomplete and terribly inefficient.

[23] In our presentation we try to reserve usage of the non-italicized "HOR" for referring to the concept of Head Order Reduction. The derivative abstract and concrete reducers are written as *HOR*, *W-HOR*, *AP-HOR*, etc. We are not certain if we have been always successful.

$$
\begin{array}{rrl}
\textit{Terms} & a,b & := & ab \,\big|\, \lambda a \,\big|\, n \\
\textit{Suspensions} & c & := & a\{\rho\} \\
\textit{Vars} & i,n & := & 0 \,\big|\, 1 \,\big|\, \ldots \\
\textit{UVCs} & \phi & := & -1 \,\big|\, -2 \,\big|\, \ldots \\
\textit{Stack} & s & := & \varnothing \,\big|\, c{:}s \,\big|\, \lambda{:}s \,\big|\, @{:}s \\
\textit{Environment} & \rho & := & \varnothing \,\big|\, c \bullet \rho \,\big|\, \phi \bullet \rho \\
\textit{Direction} & d & := & \downarrow \,\big|\, \Uparrow
\end{array}
$$

Table 3.9 The syntactic elements of the pure HOR Abstract Machine.

simple control and the natural LIFO memory usage. This property proves important when a low-level procedural casting of HOR is attempted.

The formulation of the abstract machine for HOR in the manner of Table 3.8 is patterned after the one proposed by Hilton and Berkling [Hilton 91].

HOR Narrative

Table 3.8 summarizes the workings of HOR so that they may be referred to with a quick glance. For the benefit of the reader we offer the following step-by-step account.

The workings of the *HOR* reducer are easier to comprehend if we relate them firmly to the two possible directions of traversal. The direction of traversal is the last column of the *before* and *after* portions of Table 3.8. When layers of abstraction and application are "peeled-off" we say that we are going *down* the left spine of the expression. When, starting from the head variable, the *nf* is built by piling up applications and finishing off with abstractions we say that we are going *up* the straightened spine. Complete reduction to *nf* generally entails multiple nested down-up motions; one pair for the top-level *hnf* and, recursively, one pair for each delayed subterm which emanates from the *hnf* and is collected via the stack.

First we describe what happens on the downward traversal:

Procedure down (term stack env phi)

If the term is a **variable**, it is looked up on the current environment. If it dereferences to a <u>UVC</u> environment entry then the appropriate index for the head variable implied by this constellation is calculated by subtracting the *UVC* in effect (i.e., the current ϕ) from the retrieved *UVC*. An operational *hnf* is present in the machine; each one of the machine elements contributes to its structure[24]. If the variable dereferences to a <u>suspension</u>, reduction continues with the term and the environment portion of the suspension. This last step is the one extending the left spine.

In case the term is an **application**, a suspension, constructed from the operand of the application together with the current environment, is pushed onto the stack. Traversal continues with the operator.

In case the term is an **abstraction**, the top of the stack is examined. If the top of the stack is a <u>suspension</u>, then a reduction is imminent. The necessary substitution is simulated by removing the suspension from the stack and pushing it onto the environment. We then continue with the body of the abstraction. If the top of the stack does not consist of a suspension, i.e., if the stack is <u>empty</u> or has a <u>lambda marker</u>, then a special lambda marker is pushed on the stack and the decremented *UVC* is pushed onto the environment. Traversal continues again with the body of the abstraction.

Here is what happens on the upward traversal:

Procedure up (rterm stack phi)

During upward traversal the focus is on the stack. If the stack is empty reduction is complete; the reduced term that has been built by the mutually recursive invocations of the two procedures is returned. Otherwise, the top element of the stack is examined:

If the top element of the stack is a **join marker** then the reduction of an argument has been just completed. The stack is popped twice and we continue up after making a new application with the reduced term as operand and the item on the stack that was behind the marker as an operator.

If the top element of the stack is a **lambda marker** we continue up after making a new abstraction with a fresh binder and the reduced term as body. The *UVC* is also incremented by one.

If the top element of the stack is a **suspension** we reverse direction of traversal and we continue down with the term and the environment of the suspension after pushing the reduced term and a marker onto the stack. This is done so that the completion of the reduction of a suspension on the upward traversal can be identified.

The directness and minimality of HOR may be appreciated by examining the trace of the expression *wads45* = (-2 (-1 0 (0 2)) ((-2 0 1) 0)) in Table 3.10. The reader is invited to follow this trace closely and compare it with the one for *AP* reduction in Table B.5.

---

[24] Negated, the ϕ in effect is the overall abstraction depth. Reversed, the sequence of suspensions on the stack together with the state of the environment is the spine of the *hnf*. Corrected and substituted, the UVC entry is the final head variable of the *hnf*.

step	term	stack	env	phi
D0	(-2 (-1 0 (0 2)) ((-2 0 1) 0))	nil	nil	0
D1	(-1 (-1 0 (0 2)) ((-2 0 1) 0))	(\)	(-1)	-1
D2	((-1 0 (0 2)) ((-2 0 1) 0))	(\\)	(-2 -1)	-2
D3	(-1 0 (0 2))	([((-2 0 1) 0) { -2 -1 }] \ \)	(-2 -1)	-2
•D4	(0 (0 2))	(\\)	([((-2 0 1) 0) { -2 -1 }] -2 -1)	-2
D5	0	([(0 2) { [((-2 0 1) 0) { -2 -1 }] -2 -1 }] \ \)	([((-2 0 1) 0) { -2 -1 }] -2 -1)	-2
D6	((-2 0 1) 0)	([(0 2) { [((-2 0 1) 0) { -2 -1 }] -2 -1 }] \ \)	(-2 -1)	-2
D7	(-2 0 1)	(-2 [(0 2) { [((-2 0 1) 0) { -2 -1 }] -2 -1 }] \ \)	(-2 -1)	-2
•D8	(-1 0 1)	([(0 2) { [((-2 0 1) 0) { -2 -1 }] -2 -1 }] \ \)	(-2 -2 -1)	-2
•D9	(0 1)	(\\)	([(0 2){[((-2 0 1) 0){-2 -1}] -2 -1}] -2 -2 -1)	-2
D10	0	(-2 \\)	([(0 2){[((-2 0 1) 0){-2 -1}] -2 -1}] -2 -2 -1)	-2
D11	(0 2)	(-2 \\)	([((-2 0 1) 0) { -2 -1 }] -2 -1)	-2
D12	0	(-1 -2 \\)	([((-2 0 1) 0) { -2 -1 }] -2 -1)	-2
D13	((-2 0 1) 0)	(-1 -2 \\)	(-2 -1)	-2
D14	(-2 0 1)	(-2 -1 -2 \\)	(-2 -1)	-2
•D15	(-1 0 1)	(-1 -2 \\)	(-2 -2 -1)	-2
•D16	(0 1)	(-2 \\)	(-1 -2 -2 -1)	-2
D17	0	(-2 -2 \\)	(-1 -2 -2 -1)	-2
U18	1	(-2 -2 \\)		-2
U19	(1 0)	(-2 \\)		-2
U20	(1 0 0)	(\\)		-2
U21	(-1 1 0 0)	(\)		-1
U22	(-2 1 0 0)	nil		0

Table 3.10 HOR reduction of **wads45** = (-2 (-1 0 (0 2))((-2 0 1) 0)).

Strong HOR was first suggested by Berkling via his presentation of the in-the-large rewrite rules (3.2)-(3.4) in [Berkling 87]. A concrete "in-the-small" formulation of the core of HOR and of an architecture having an applied Lambda Calculus as its machine language is the subject of US patent[25] #5,099,450. Berkling coined the name "Head Order" because reduction is effected only on the redices along the left spine of the overall term and traversal proceeds to the head variable of an expression. Hilton develops formalizations of HOR for an applied Lambda Calculus in [Hilton 90b].

With our parting words for this chapter we would like to reiterate that the rules for *HOR* in Table 3.8 describe an abstract machine which does not rely on recursive calls to itself. This may be verified by studying the abstract machine and by examining the concrete description in Figure A.9 of Appendix A and by noting that all calls to *down* and *up* are tail recursive. Because the reversed spine or "stack" is explicitly present, a simple automaton with iterative control suffices for implementation. This endows *HOR* the same level of "implementability" as that of the *SECD* machine even though it follows a normal order strategy and advances to strong normal form. In contrast, the *AP* reducer as described in

---

[25] Filed 22 Sep. 1988, awarded 24 Mar. 1992; Inventor: Klaus J. Berkling; Assignee: Syracuse University.

section 3.5 uses non-tail calls to itself which implies that a low-level implementation will require the addition of at least one stack. In this sense *AP* is more like *eval-apply*.

Chapter 4

Facets of Head Order Reduction

4.1 HEAD-NORMAL REDUCTION

In this chapter we focus on the operational details of Head Order Reduction. First, the construction of the lazy *operational head normal form* (*ohnf*) is further explicated. These *ohnf*s are calculated by a single down-up traversal of the expression graph without processing the argument suspensions. A vital optimization which makes single-variable operands much more efficient is discussed next. After sketching the reasons that assure us of correctness, we describe a modified sequencing for HOR which partially abandons depth-first traversal. Finally, mechanisms of incremental reduction and expression navigation that rely on the HOR machinery are presented.

The workings of pure HOR are actually a superset of the machinery needed when reduction stops at *hnf*. Table 4.1 shows an abstract machine that is a slightly simplified version of that of section 3.6. The only difference is in the rule for a suspension encountered on the upward traversal. Instead of interrupting the traversal to reduce it in isolation, the suspension (shown underlined) is regarded as a delayed operand and is included as is. All such pending suspensions share the common abstraction context of the *hnf* whose formation is complete when there are no more elements on the stack.

Recapitulating, the *HN-HOR* reducer reduces an expression partially to an operational head normal form via a single down-up traversal by incorporating pending suspensions as bona fide arguments. In terms of the pictures of the previous chapter it performs the "straightening" of the left spine producing a single east-pointing corner.

Term	Stack	Env	UVC	Dir	Term	Stack	Env	UVC	Dir
ab	s	ρ	ϕ	$\downarrow$	a	$b\{\rho\}{:}s$	ρ	ϕ	$\downarrow$
λa	$c{:}s$	ρ	ϕ	$\downarrow$	a	s	$c\bullet\rho$	ϕ	$\downarrow$
λa	s	ρ	ϕ	$\downarrow$	a	$\lambda{:}s$	$\phi-1\bullet\rho$	$\phi-1$	$\downarrow$
$i+1$	s	$u\bullet\rho$	ϕ	$\downarrow$	i	s	ρ	ϕ	$\downarrow$
0	s	$a\{\rho'\}\bullet\rho$	ϕ	$\downarrow$	a	s	ρ'	ϕ	$\downarrow$
0	s	$\phi'\bullet\rho$	ϕ	$\downarrow$	$\phi'-\phi$	s	$-$	ϕ	$\uparrow$
a	$\lambda{:}s$	$-$	ϕ	$\uparrow$	λa	s	$-$	$\phi+1$	$\uparrow$
a	$b\{\rho'\}{:}s$	$-$	ϕ	$\uparrow$	$ab\{\rho'\}$	s	$-$	ϕ	$\uparrow$
a	$\{\}$	$-$	ϕ	$\uparrow$	a	s	$-$	$-$	STOP

Table 4.1 Head-Normal HOR Abstract Machine.

Depictions of Lazy Head Normal Forms

The operational head normal form is interesting in its own right. The suspensions that are pending to the right of the spine have a special structure as predicted by the in-the-large rules of *HOR*.

We start by giving a concrete example of the matched abstraction-argument pairs which form an environment that is prepended to selected subterms of the original problem expression. Figure 4.1 (*B*) shows the operational *hnf* of ((-1 0 0) (-2 (-1 0 (1 0)) ((-1 0 1 0) 1))) as two interconnected trees corresponding to the nested triangle picture shown in section 3.6.

After the down-up traversal is complete, the original spine (0) is not reachable any more. It has been superseded by the straight vertical line on the left of the *hnf* tree. The suspensions are formed lazily by prepending portions of the original expression with the proper environments symbolized by the nested abstraction-suspension/*UVC* terms.

On the next page (Figure 4.2) we show the corresponding graph that is left in the reduction machine after traversal. It consists of implementation-level nodes—i.e., SUSP and LINK nodes in addition to the ordinary abstraction and application ones. The one-to-one correspondence of this graph to the nested environment via lambda expressions of Figure 4.1 (*B*) is an assurance that the algorithm employed has indeed produced the correct result.

To underline our options in depicting *hnfs* the graph, *BTF* and *Left Spine* formats of the previous *hnf* are shown in Table B.11 (Appendix B). The graph format is useful for showing the size and internal sharing of the result. The left spine format—utilized in the interactive reduction display of our experimental system covered in Appendix C—is useful when one prefers to concentrate on the spine.

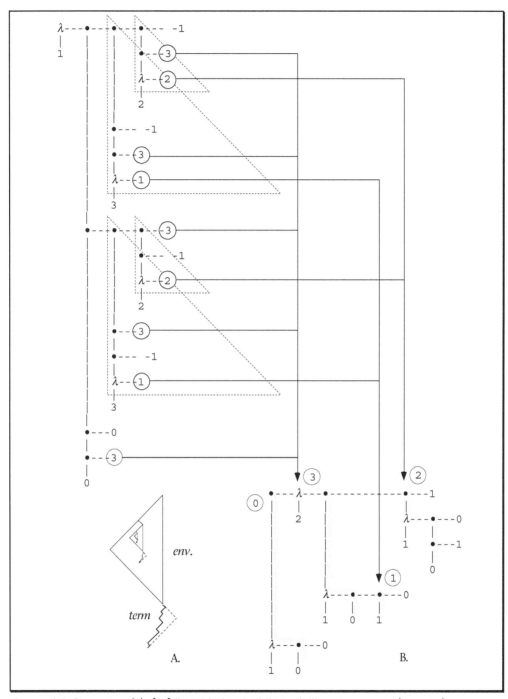

Figure 4.1 Operational hnf of ((-1 0 0) (-2 (-1 0 (1 0)) ((-1 0 1 0) 1)))   via recursively nested environment triangles.

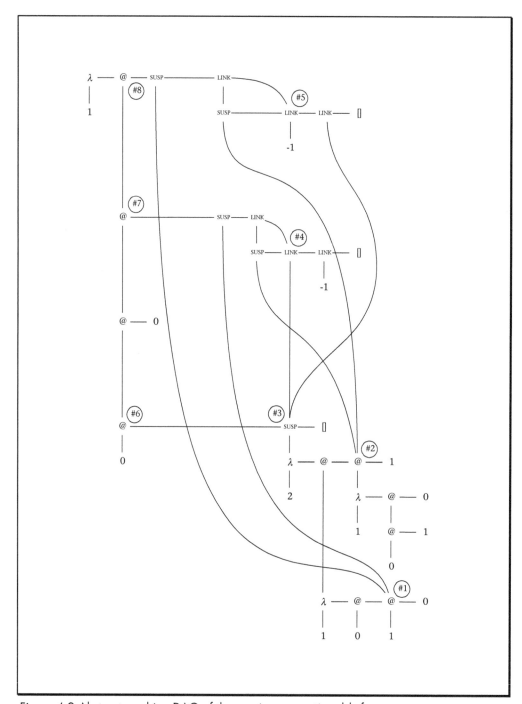

Figure 4.2 Abstract machine DAG of the previous operational hnf.

The left spine format consists of writing explicit application nodes on a vertical line, the overall abstraction count on the top left, and the head variable at the bottom. This format is also a prelude to the low-level linear memory representation that is suited to *HOR*.

Finally, for the really curious reader, we show in Table B.9 the full sequence of steps leading to this operational *hnf*. It is given in the form of a trace of the machine states that produces it. Displaying all of these intermediate steps of the head-normal HOR reduction (*HN-HOR*) in the linear *BTF* format is too cumbersome; therefore they are shown in the raw internal graph form so that the sharing of structure remains visible. Each row of the middle column in this table consists of the term, stack and environment structures separated by vertical bars. As with all other traces, the leftmost column counts the iterations, tagging them with the direction of traversal and additionally, with a solid dot whenever a reduction has been effected. Graph information is preserved by annotating a shared subgraph G with a unique integer n (i.e., #n=G) and by referring to all occurrences of it after the leftmost one by a special symbol (i.e., #n#).

4.2 ASSURANCES OF CORRECTNESS FOR HOR

Zhang and Berkling have established in [Zhang 89] that (*a*) the rules of HOR in-the-large generate terms that preserve equivalency and (*b*) that, if an expression has one, then application of these rules will terminate with its head normal form. Via a simple argument this result is extended to full normal forms.

For the most part, the detailed "in-the-small" state transitions of HOR presented in section 3.6 are easy to justify. On one hand their formulation follows quite easily from the in-the-large rules and on the other they appeal to intuition since they have great affinity to the state transitions of other minimal functional interpreters (cf. *EV-AP* and *SECD*). But transparency is not nearly as obvious in the case of the "ϕ-updating" operation of a head variable which dereferences to a *UVC* entry[26]. We therefore offer the following picture and explanation. A more formal argument is included in section 6.3.

Figure 4.3 shows a snapshot of HOR reduction of a subterm after *hnf* has been reached. An operand consisting of a sole variable v, is looked up on the applicable environment— shown as a triangle that has collected all the delayed substitutions of the parent *hnf*—and it happens to dereference to a *UVC* entry. The parent *hnf* has a abstraction-depth context of $-n$. Now, the *UVC* entry was made when a fresh unbound variable was introduced via an η-extension which was effected because an unapplied abstraction was traversed. The specific value $-m$, of this *UVC* entry signifies that at the point of the introduction this variable was

---

[26] This head variable can be the head of the overall *hnf* or of a lone variable operand. We show the latter case.

the mth η variable introduced in the reduction of this subterm. Hence, the correct de Bruijn index for the variable v is one which establishes this association, namely $n - m$. Since the hnf is guaranteed not to undergo any further structural changes, this value is also the final one.

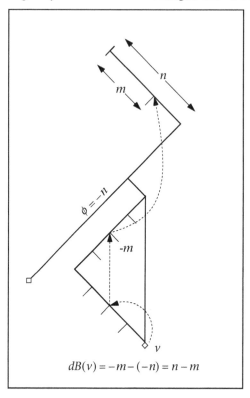

Figure 4.3 Calculating the correct dB index of an uninstantiated variable.

4.3 SINGLE VARIABLE ARGUMENTS

The opportunity for an important optimization of *HOR* manifests itself in the case of an application whose argument is a single variable. Instead of creating a closure with the sole purpose of looking up its variable term when, and if in fact, it becomes a focus expression, the variable argument is dereferenced immediately and the looked-up expression is pushed onto the result stack. Note that since a looked-up value can now be a *UVC*, during the upward motion of the reduction mechanism a *UVC* encountered at the top of the stack must be ϕ-updated before it becomes the argument variable belonging to the final result.

This augmentation is termed *Early Variable Argument Lookup* or simply *VA*. The abstract machine descriptions of *HOR* are modified as described next.

It is easy to add early *VA* in the context of the weak *HOR* reducer. A variable argument is simply looked-up instead of becoming the trivial term of a closure. The fist rule of Table 4.2 shows this direct dereferencing. One should be aware that the rules are no longer disjoint; therefore the order of their application is important.

Term	Stack	Env	Term	Stack	Env
ai	s	$u_0 \bullet \ldots \bullet u_i \bullet \ldots \bullet \rho$	a	$u_i{:}s$	$u_0 \bullet \ldots \bullet u_i \bullet \ldots \bullet \rho$
ab	s	ρ	a	$b\{\rho\}{:}s$	ρ

Table 4.2 Additional rules of early VA lookup for W-HOR.

Under strong *HOR*, early *VA* look-up is defined by the rewrite rules of Table 4.3.

Term	Stack	Env	UVC	Dir	Term	Stack	Env	UVC	Dir
ai	s	$u_0 \bullet \ldots \bullet u_i \bullet \ldots \bullet \rho$	ϕ	$\downarrow$	a	$u_i{:}s$	$u_0 \bullet \ldots \bullet u_i \bullet \ldots \bullet \rho$	ϕ	$\downarrow$
0	s	$\phi' \bullet \rho$	ϕ	$\downarrow$	$\phi' - \phi$	s		ϕ	$\uparrow$
a	$\phi'{:}s$	—	ϕ	$\uparrow$	$a(\phi' - \phi)$	s	—	ϕ	$\uparrow$

Table 4.3 Additional rules of early VA lookup for HOR.

The syntax rules of Table 3.9 must also be extended so that the stack can have *UVC* indices. These ϕ entries can participate in β-reductions:

$$Stack \quad s := \varnothing \,|\, c{:}s \,|\, \phi{:}s \,|\, \lambda{:}s \,|\, @{:}s \tag{4.1}$$

One repercussion of early *VA* under strong *HOR* is that a *UVC* index can now become an environment element not only directly—as the result of traversing an unapplied λ—but also indirectly as the result of a reduction. In section 6.3 we will ascertain that this shortcut, although beneficial and completely innocent in the context of the pure *HOR*, leads to a major complication in the sharing version of *HOR*. Subtle errors are circumvented by effecting a selective correcting traversal when the binding context of creation is different from the binding context of an argument occurrence.

Table B.12 shows the differences in the linear graph and *BTF* representations of the operational *hnf*'s for the previous test expression with both early and late *VA* lookups. Typically, an *ohnf* resulting from early *VA* lookup is more compact than one without it. The differences are much more pronounced after unraveling to *BTF* format.

Other, relatively self-contained, optimizations of *HOR* are discussed after the introduction of sharing. In the next section we present a variation which adheres to the overall principles of *HOR* but partly abolishes the depth-first eager traversal. This version is proposed by Berkling as a stepping stone towards the quest for a fully lazy applied reducer.

4.4 Breadth-First HOR Strategy

To motivate the creation of the breadth-first *HOR*, we need to remind ourselves of the structural invariance of a *hnf* under reduction. When substitutions are fully effected this property of a *hnf* is quite obvious. It is more difficult to see that the same is true when substitutions are delayed. An important observation under both scenarios is that since the subterms of an *hnf* can refer to top-level abstractions, one has to reduce the subterms comprising the *hnf* in the proper abstraction context. This context is uniquely determined by the count of the top-level abstractions of the parent *hnf*.

e	((-1 0 0) (-2 (-1 0 (1 0)) ((-1 0 1 0) 1))) $\rightarrow_8$
h	(-1 $\underline{0}$ (-2 (-1 0 (1 0)) ((-1 0 1 0) 1)) $\underline{0}$ ((-2 (-1 0 (1 0)) ((-1 0 1 0) 1)) ((-1 0 (-2 (-1 0 (1 0)) ((-1 0 1 0) 1)) 0) 0)) (0 ((-1 0 1 0) (-2 (-1 0 (1 0)) ((-1 0 1 0) 1))))) = (-1 0 *A B C D*)

Table 4.4 Head-normal reduction of e to its head normal form, h.

A.	(-2 (-1 0 (1 0)) ((-1 0 1 0) 1)) $\rightarrow_2$ (-2 1 0 1 (0 (1 0 1)))
B.	$\underline{0} \rightarrow_0 \underline{0}$
C.	((-2 (-1 0 (1 0)) ((-1 0 1 0) 1)) ((-1 0 (-2 (-1 0 (1 0)) ((-1 0 1 0) 1)) 0) 0)) $\rightarrow_{20}$ (-1 1 (-2 1 0 1 (0 (1 0 1))) 1 0 (1 (-2 1 0 1 (0 (1 0 1))) 1) (0 (1 (-2 1 0 1 (0 (1 0 1))) 1 0 (1 (-2 1 0 1 (0 (1 0 1))) 1))))
D.	(0 ((-1 0 1 0) (-2 (-1 0 (1 0)) ((-1 0 1 0) 1)))) $\rightarrow_{28}$ (0 (0 (-2 1 0 1 (0 (1 0 1))) 0 (-1 1 (-2 1 0 1 (0 (1 0 1))) 1 0 (1 (-2 1 0 1 (0 (1 0 1))) 1) (0 (1 (-2 1 0 1 (0 (1 0 1))) 1 0 (1 (-2 1 0 1 (0 (1 0 1))) 1))))))

Table 4.5 Normal forms of the arguments emanating from the spine of h.

nf	(-1 $\underline{0}$ (-2 1 0 1 (0 (1 0 1))) 0 (-1 1 (-2 1 0 1 (0 (1 0 1))) 1 0 (1 (-2 1 0 1 (0 (1 0 1))) 1) (0 (1 (-2 1 0 1 (0 (1 0 1))) 1 0 (1 (-2 1 0 1 (0 (1 0 1))) 1)))) (0 (0 (-2 1 0 1 (0 (1 0 1))) 0 (-1 1 (-2 1 0 1 (0 (1 0 1))) 1 0 (1 (-2 1 0 1 (0 (1 0 1))) 1) (0 (1 (-2 1 0 1 (0 (1 0 1))) 1 0 (1 (-2 1 0 1 (0 (1 0 1))) 1)))))))

Table 4.6 Overall normal form of e = ((-1 0 0)(-2 (-1 0 (1 0))(-1 0 1 0) 1))).

Consider the reduction sequence of Table 4.4 which shows a head normal reduction of the expression *tw7-2*. The *hnf* of this expression, reached after eight steps, consists of four subterms *A – D*, in addition to the outer abstraction and the head variable. In Table 4.5 each of the argument subterms has been fully reduced in the abstraction context of a single outer abstraction.

Table 4.6 lists the overall *nf* of *tw7-2* which can indeed be produced by the literal substitution of each of the reduced subterms in the *hnf* (-1 0 *A B C D*). These subterms can be reduced independently of each other.

Breadth-First Traversal

We show now how a modified order of traversal can take advantage of the invariance of the *hnf* under the delayed model of HOR.

The expression is traversed performing the standard HOR processing but without visiting the argument suspensions during the ascent of the straightened left spine. Instead, the application nodes are left having these suspensions as their operands. These application nodes are collected in a set. Each suspension is annotated with the ϕ which is in effect[27]. The suspensions of each *hnf* are added to the set as they are encountered. When the ascent of the skeleton *hnf* is completed, the reduction mechanism is called recursively on the oldest delayed suspension. When one of the suspensions has been processed in this piecemeal manner, the application node in the parent skeleton is updated to point to the resulting *hnf*. The set of the delayed suspensions is referred to as the *process queue* due to the FIFO discipline which is obeyed. The whole process is guaranteed to terminate if the expression has an overall normal form.

A picture is helpful in visualizing this strategy. Figure 4.4 (*A*) shows that a lambda expression may be visualized as a polygon with a zig-zag line on its west side symbolizing the expression's left spine and a right angle corner on the east side symbolizing the resulting operational *hnf*. The straightened spine of the *hnf* consists of application nodes with pending suspensions as operands. With this picture in mind, the *BF-HOR* order can then be defined by the breadth-first traversal of the pending suspensions of each *hnf* as in part (*B*). Pure *HOR* uses the eager depth-first traversal shown for reference in part (*C*).

A more formal definition of this algorithm is not given since the core of the processing is that of *HN-HOR* augmented with the mechanism for keeping a record of all the delayed suspensions (cf. section 4.1).

The top-level control of *BF-HOR* is defined in terms of a function which, given a top-level *hnf* and a process queue, repeatedly invokes the augmented *HN-HOR* algorithm on each element of the queue and updates the parent spine after each *down-up* traversal is done. The overall process has to be properly started with a pseudo *hnf* and a single element queue which includes the problem term. A concrete *BF-HOR* reducer appears in Figure A.12.

The *BF-HOR* strategy seems at first quite more elaborate than that of pure *HOR*. The fact that each delayed suspension is refined only slightly may make one think that the process has appreciably higher memory requirements than *HOR*. But a closer look reveals that the overhead of *BF-HOR* is rather minimal. The partial *hnf*s are mostly "glue" and most of their structure is shared with existing portions of the graph. Each in-situ updating of a delayed suspension is in an incremental refinement which becomes part of the final result.

---

[27] Berkling uses the term *superclosure* for this package.

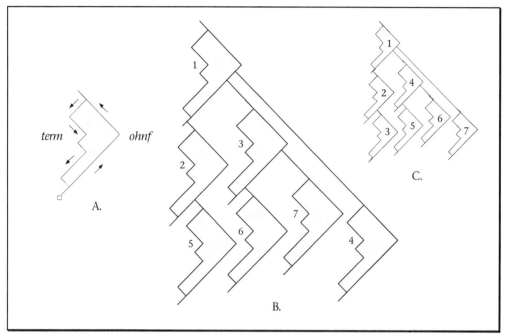

Figure 4.4 Order of traversal for BF-HOR.

Regarding memory utilization we would like to make an additional point. An implementation of pure *HOR* can use an organization of the environment which offers the possibility of discarding unused portions when the reduction of a suspension is completed. Under the *BF-HOR* strategy the reduction of arguments is incremental and therefore the environment structure is needed until the very end. We have not yet determined how this affects the overall memory usage but limited evaluation attests that the increase is indeed modest for most expressions.

Successive hnfs under BF-HOR

We now illustrate the *BF-HOR* processing with a detailed example. Consider the expression (-2 (-1 (-1 0 (0 1)) (-1 2 0 1)) (-1 0 2)), listed in tree form in Figure 4.5.

Figure 4.6 shows the top-level skeleton and the process queue at each stage after successive suspensions have been processed. The top-level *hnf*, shown in part (*A*) is reached after three reductions. Traversal results in a process queue of two elements, each one of which points to the two application nodes of the top-level *hnf*.

After processing of the first suspension of the process queue the situation is as in part (*B*). The first element of the queue has been removed and two new ones have been added pointing to the application nodes of the newly created operand *hnf*. All three elements of the process queue point to application nodes which share the same operand suspension. Why this is the case is left as an exercise for the reader. Also note that all three share the same abstraction context since the newly formed *hnf* does not have any new abstractions.

The rest of the traversals are straightforward. In each case the shared suspension [(-1 0 2) { -2 -1 }] is trivially transformed to (-1 0 2) and no further elements are added to the process queue. The successively updated top-level *hnf* is shown in parts (*C*) through (*E*). A complete trace is shown in Table B.10.

We conclude the discussion of this important variation of *HOR* with a last observation. In step (*B*) of the sequence above, we see that the three elements of the process queue actually refer to the same suspension. Therefore, the issue arises whether sharing can be accomplished via the process queue. When processing of a suspension is complete, instead of updating the application node to point to the newly created *hnf,* one could update the original cell of the suspension so that any further occurrences of the same suspension in the process queue can take advantage of the ready *hnf.* This may be done safely if other occurrences of the target suspension in the queue are scheduled to be processed with an identical ϕ. In general this is not the case though. A more elaborate mechanism is needed which tags the result with the ϕ under which it was processed. This mechanism is also required when the pure *HOR* is augmented with sharing. It will be described in detail in section 6.3.

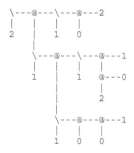

Figure 4.5 Tree form of tw10 = (-2 (-1 (-1 0 (0 1))(-1 2 0 1))(-1 0 2)).

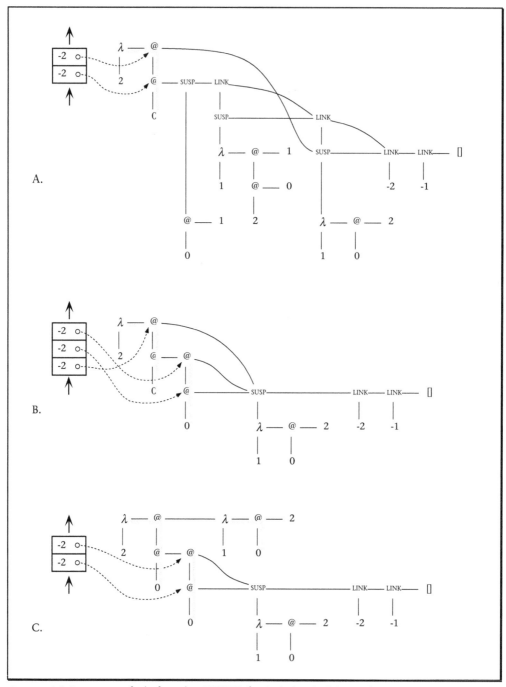

Figure 4.6 Sequence of ohnfs under BF-HOR for (-2 (-1 (-1 0 (0 1)) (-1 2 0 1)) (-1 0 2) .

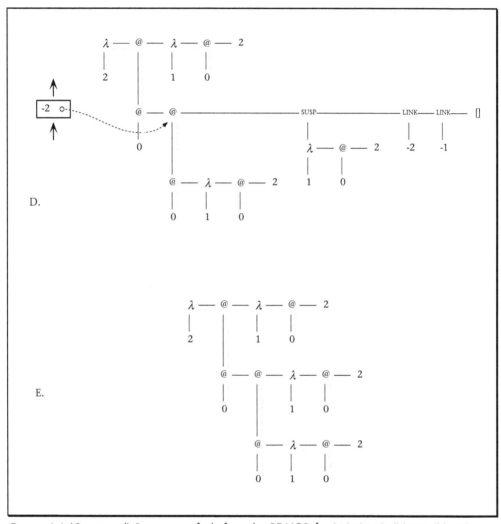

Figure 4.6 (Continued) Sequence of ohnfs under BF-HOR for (-2 (-1 (-1 0 (0 1)) (-1 2 0 1)) (-1 0 2)) .

4.5 INCREMENTAL REDUCTION AND EXPRESSION NAVIGATION

Controlling Reduction with a Quantum

Incremental reduction is the ability to limit the number of contractions performed on a target expression. Under the procedural model of computing one starts a program by giving a "run" command. Under a reduction-based model one can include with the "reduce" command a *quantum* whose value represents the maximum allowable number of reductions to be performed on a target expression. Controlling reduction is such a manner can keep under check the processing of expressions that do not have a normal form. Incremental reduction also permits us to observe the symbolic execution of a program step by step.

Term	Stack	Env	UVC	Dir	Quan	Term	Stack	Env	UVC	Dir	Quan
λa	$c{:}s$	ρ	ϕ	$\downarrow$	0	a	$\lambda{:}c{:}s$	$\phi - 1 \bullet \rho$	$\phi - 1$	$\downarrow$	0
λa	$c{:}s$	ρ	ϕ	$\downarrow$	$q+1$	a	s	$c \bullet \rho$	ϕ	$\downarrow$	q

Table 4.7 Augmentation of the HOR Abstract Machine with a reduction quantum.

The mechanisms of *HOR* are an excellent match to incremental reduction. The focus on the left-spine and the η-extensions effected by the *UVC* mechanism make it exceedingly simple to achieve incremental reduction. The necessary changes in the abstract machine of *HOR* (Table 3.8) consist of the addition of an integer quantum and the splitting of the rule for β-contraction in two parts.

Under the new rules (shown in Table 4.7) a contraction is allowed to happen only if the quantum is not exhausted. For each contraction the quantum is decremented. In case the quantum is expired the redex is preserved by effectively treating the abstraction as an unapplied one. A lambda marker is pushed on the stack and the effective *UVC* is updated to reflect the traversing of the abstraction. During upward traversal the ordinary processing of *HOR* reconstructs the partially reduced expression and the denied redices remain intact in the overall result.

Limiting reduction via the quantum has many applications. It can help with debugging and it allows one to see the effect of sharing by freezing the parallel substitutions that sharing permits[28]. If the quantum is set to one before each invocation of the reducer the effect is that of single-stepping.

---

[28] Sharing under HOR is introduced in section 6.3. It is recommended that it is reviewed before proceeding.

0.	((-1 0 0) ((-2 1 1) (-1 0) (-1 0)))
1.	((-2 1 1) (-1 0) (-1 0) ((-2 1 1) (-1 0) (-1 0)))
2.	((-1 (-1 0) (-1 0)) (-1 0) ((-2 1 1) (-1 0) (-1 0)))
3.	((-1 0) (-1 0) ((-2 1 1) (-1 0) (-1 0)))
4.	((-1 0) ((-2 1 1) (-1 0) (-1 0)))
5.	((-2 1 1) (-1 0) (-1 0))
6.	((-1 (-1 0) (-1 0)) (-1 0))
7.	((-1 0) (-1 0))
8.	(-1 0)

Table 4.8 Quantum-controlled HOR reduction of ((-1 0 0)((-2 1 1)(-1 0)(-1 0))).

0.	((-1 0 0) ((-2 1 1) (-1 0) (-1 0)))
1.	((-2 1 1) (-1 0) (-1 0) ((-2 1 1)...(-1 0)...(-1 0)))
2.	((-1 (-1 0) (-1 0)) (-1 0) ((-1 (-1 0) (-1 0))...(-1 0)))
3.	((-1 0) (-1 0) ((-1 0)...(-1 0)))
4.	((-1 0) (-1 0))
5.	(-1 0)

Table 4.9 Quantum-controlled HOR-SH reduction of ((-1 0 0)((-2 1 1)(-1 0)(-1 0))).

Tables 4.8 and 4.9 show incremental reduction of Wadsworth's example expression under *HOR* and *HOR-SH* respectively. Under the sharing strategy three redices are reduced in parallel. They are highlighted with dotted underlining. Note that in the case of sharing it is not sufficient to set the quantum to one and make the target expression at each step the result of the previous one. When the quantum is set to one the incremental behavior of the sharing versions is indistinguishable from that of the non-sharing ones. The quantum has to be set to each of the values listed on the left column starting every time with the original expression. If only a single reduction is allowed there is no opportunity for reduction in isolation to be completed and for its outcome to have an effect.

Table 4.10 shows a detailed trace of *HOR* reduction with sharing in which one can observe that the last reduction happens at step *D15* when the quantum reaches zero. The abstraction at step *D17* is not permitted to participate in a β-contraction. Instead, a copy of it is made—as it should, since it is an operator—during steps *U19-20*.

step	term/shexp	stack	env	phi	q
D0	((-1 0 0) ((-2 1 1) (-1 0) (-1 0)))	nil	nil	0	4
D1	(-1 0 0)	((([((-2 1 1) (-1 0) (-1 0)) ({ })]))	nil	0	4
•D2	(0 0)	nil	((([((-2 1 1) (-1 0) (-1 0)) ({ })]))	0	3
D3	0	((([((-2 1 1) (-1 0) (-1 0)) ({ })]))	((([((-2 1 1) (-1 0) (-1 0)) ({ })]))	0	3
•D4	((-2 1 1) (-1 0) (-1 0))	nil	nil	0	3
D5	((-2 1 1) (-1 0))	((([(-1 0) ({ })]))	nil	0	3
D6	(-2 1 1)	((([(-1 0) ({ })]) ([(-1 0) ({ })]))	nil	0	3
•D7	(-1 1 1)	((([(-1 0) ({ })]))	((([(-1 0) ({ })]))	0	2
•D8	(1 1)	nil	((([(-1 0) ({ })]) ([(-1 0) ({ })]))	0	1
D9	1	((([(-1 0) ({ })]))	((([(-1 0) ({ })]) ([(-1 0) ({ })]))	0	1
D10	(-1 0)	nil	nil	0	1
D11	0	(\)	(-1)	-1	1
U12	0	(\)		-1	1
U13	(-1 0)	nil		0	1
R10	(0.0 -1 0)	.	.		
D14	(-1 0)	((0.0 -1 0))	nil	0	1
•D15	0	nil	((0.0 -1 0))	0	0
U16	(-1 0)	nil		0	0
R4	(0.0 -1 0)	.	.		
D17	(-1 0)	((0.0 -1 0))	nil	0	0
D18	0	(\ (0.0 -1 0))	(-1)	-1	0
U19	0	(\ (0.0 -1 0))		-1	0
U20	(-1 0) *copy	((0.0 -1 0))		0	0
U21	((-1 0) (-1 0))	nil		0	0

Table 4.10 Quantum-controlled HOR-SH-VA reduction of ((-1 0 0)((-2 1 1)(-1 0)(-1 0))).

Incremental reduction as a native mechanism for controlling reduction is not available directly under alternative strategies which do not strictly descend the left spine. In the *RTNF/RTLF* strategy, for example, limiting reductions with a quantum results in expressions which are not terms of the pure Lambda Calculus but which include delayed substitutions in the form of suspensions. As usual, *BTF* proves be a good notation for displaying such mixed terms (Table 4.11).

0.	(-2 (-1 0 (0 2)) ((-2 0 1) 0))
1.	(-2 ([(-2 0 1) ({ 0 1 })]) 0 (([(-2 0 1) ({ 0 1 })]) 0 1))
2.	(-2 ([(-1 0 1) ({ ([0 ({ 0 1 })]) 0 1 })]) (([(-2 0 1) ({ 0 1 })]) 0 1))
3.	(-2 ([(-2 0 1) ({ 0 1 })]) 0 1 0)
4.	(-2 ([(-1 0 1) ({ ([0 ({ 0 1 })]) 0 1 })]) 1 0)
5.	(-2 1 0 0)

Table 4.11 Quantum-controlled reductions of (-2 (-1 0 (0 2))((-2 0 1) 0)) under AP.

We make one additional point of a more general nature. Reduction with a quantum reinforces our credo about the relative costs of reduction. Since the number of reductions are under tight control, we see that the processing costs are dominated by the costs of traversal. Incremental reduction with the quantum set to one or zero often exhibits a cost comparable to that of a single "free-rolling" traversal which takes an expression to its normal form[29]. Under incremental reduction the requisite mechanisms have to be set up as if all reductions are to be performed. The amortization of these setup costs over many reductions is what makes full reduction efficient.

Expression Navigation

The mechanisms of *HOR* also provide a remarkably direct solution to the problem of "navigating" an expression by varying the focus of attention. They make it possible to selectively reduce the focus subterm in an equivalence-preserving manner without any limitations. In an applied reduction system this property translates to the ability to interactively establish any portion of a program as a focus and subsequently reduce it in situ as much as it is desired.

Term	Stack	Env	UVC	Dir	Term	Stack	Env	UVC
ab	s	ρ	ϕ	$\downarrow$	a	$b{:}s$	ρ	ϕ
ab	s	ρ	ϕ	$\rightarrow$	b	$@{:}a{:}s$	ρ	ϕ
λa	s	ρ	ϕ	$\downarrow$	a	$\lambda{:}s$	$\phi-1\bullet\rho$	$\phi-1$
a	$\lambda{:}s$	$\phi-1\bullet\rho$	$\phi-1$	$\uparrow$	λa	s	ρ	ϕ
b	$@{:}a{:}s$	ρ	ϕ	$\leftarrow$	ab	s	ρ	ϕ
a	$b{:}s$	ρ	ϕ	$\uparrow$	ab	s	ρ	ϕ

Table 4.12 Equivalence-preserving lambda expression navigation.

Dir.	Focus is an:	New Focus is the:
$\downarrow$	application	application's operator
$\downarrow$	abstraction	abstraction's body
$\rightarrow$	application	application's operand
$\leftarrow$	operand	operand's parent application
$\uparrow$	abstraction's body	abstraction
$\uparrow$	operator	operator's parent application

Table 4.13 Definition of the effect of a move of focus within an expression.

Table 4.12 shows the rules that make navigation possible. The meaning of the symbol showing the direction of the move in the middle column is summarized in Table 4.13. These rules define the traversal of an expression by effecting η-extensions via the *UVC* mechanism when abstractions are traversed and by pushing skipped operands onto the stack. At all times during traversal the state of the machine constitutes necessary and sufficient information to

---

[29] This is true of expressions without self-application which can lead to an unbounded amount of work. In the extreme, an expression like $\Delta = (\lambda x.xx)(\lambda x.xx)$ requires infinite resources for "full" reduction while incurring only a constant "per reduction" cost.

reconstruct the parent expression. The reduction machinery may be called at any point during this traversal and its results are to be interpreted under the context created by the application of the rules of Table 4.12.

Given our underlying assumption of globally closed *wfes*, the initial expression on which traversal is started must be an abstraction. Hence, the initial state is constrained to be a quadruplet of the form $\langle \lambda a, \varnothing, \varnothing, 0 \rangle$. The only possible choice of the first move is that of the third row of Table 4.12.

Expression navigation has been a prominent feature of Berkling's reduction systems since the early 1970's. It is usually combined with syntax-directed editing [Hommes 78].

4.6 RTNF/RTLF STRATEGY WITH HOR MACHINERY

Having covered the *RTNF/RTLF* and *HOR* strategies and the usual representation decisions with which they are usually associated an important question surfaces: Can they be decoupled? Specifically, is it possible to implement the *RTNF/RTLF* strategy using HOR machinery and the HOR strategy using pointer variables? Answering this question can provide us with additional insights because it allows one to gauge the effect of strategy separately from the effects of representation.

We have succeeded in answering affirmatively both parts of the last question. Figures A.7 and A.10 show respectively a HOR reducer utilizing REP-style structures and an *AP* reducer with a HOR-style machinery. Since we have already covered the two strategies we concentrate below on the implementation details. The reader is advised that the paragraph which covers the particulars of sharing of the hybrid *AP-HOR* reducer requires a review of section 6.3.

First we discuss the interesting details of the *AP-HOR* reducer. The HOR-style mechanism consists of the inclusion of *UVC* environment entries and the ϕ register for keeping track of varying abstraction contexts. The *VA* optimization which was devised for HOR is also directly applicable. Other than the inclusion of the ϕ register in all the recursive invocations, *AP-HOR* follows directly the structure of *AP*.

Only when the term is a variable the differences are worth mentioning. First, environment lookup when HOR structures are used is a indexing operation based on the de Bruijn index of the variable term instead of a sequential searching operation that tries to match variable pointers. Second, when a *UVC* entry is retrieved in head position the ordinary ϕ-correction is effected. The remaining differences of variable processing have to do with sharing.

Sharing under *AP-HOR* is patterned exactly after the one of the Aiello-Prini reducer. When the isolated reduction of an operator suspension results in a another suspension, the new suspension updates the one that initiated the isolated reduction. If reduction in isolation returns a *shexp*, the *shexp* replaces the suspension that initiated the isolated reduction. When a variable lookup dereferences to a *shexp* which was computed under an abstraction context different from the current one, the *shexp* is reentered with the proper "dummy" environment. As in the case of *HOR-SH* abstraction context tagging for *shexps*, may lead to extra traversals when a correction of de Bruijn indices is required. As in the case of *HOR-SH*, *shexps* are introduced into the environment both as a result of in-situ updating and a result of early *VA* and subsequent reduction.

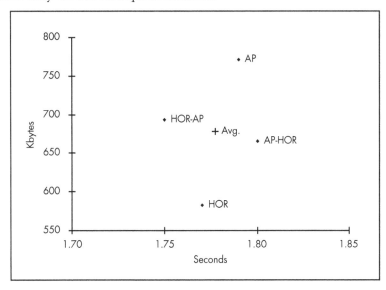

Figure 4.7 Memory and time for the all* suite under four pure normal-order reducers.

AP-HOR is more economical in terms of memory than *AP* because environment entries need only be values not variable-pointer value pairs. Another, but statistically not as significant, source of savings is the abolition of the nonessential nodes which are the targets of the pointer variables. The validity of these remarks is substantiated by measurements which compare the time and memory requirements of these four normal order reducers on the test set we will be using in chapter 7. Figure 4.7 plots these results.

One should comment on the remarkable similarity of the runtime statistics of these reducers. All four are within ±1.5% of the average in terms of time and within ±14% of the average in terms of memory. This must be an indication that these reducers "get to the heart of the matter". The space advantages of the HOR machinery are judged as significant given the overall minimality of these reducers. In a low-level "machine" realization the tail-recursive nature of *HOR* is going be a further benefit.

HOR strategy under a REP-style

The casting of the HOR strategy with REP-style structures is entirely straightforward. Instead of the ϕ register and the *UVCs*, globally new pointer variables are introduced for each unapplied abstraction. During the ascent of the spine these new pointer variables become the binders the outer abstractions of the operational head normal form. Figure A.10 lists a concrete *"HOR-AP"* reducer with early *VA* optimization.

We have not solved the problem of whether sharing in the example of *HOR-SH* (cf. section 6.3) can be effectively added to this hybrid reducer.

Chapter 5

Related Theories

5.1 DE BRUIJN'S LAMBDA CALCULUS

We devote now a few pages in outlining the relation of the techniques of the previous two chapters to de Bruijn's Lambda Calculus and to two recent theories that treat substitutions formally. We point out the similarities and show that the machinery of HOR can be considered as a distillation of these theories when the focus is efficient strong normalization. The similarities are notable given that HOR techniques were developed by Berkling independently and before the more recent theories were formulated.

In his classic paper N.G. de Bruijn [Bruijn 72] shows that the names used to denote occurrences of bound variables in the Lambda Calculus are not really necessary. A notation by which the names are replaced by indices counting the "distance" between bound variable occurrences and their respective binders suffices for preserving meaning. It also facilitates automatic formula manipulation because during substitution with named variables a systematic renaming of variables is necessary in order to avoid the situation of having a free variable being inadvertently captured by an abstraction.

The following example expression with parenthesized prefix function symbols a, b, and f is utilized by de Bruijn to illustrate the technique.

$$\lambda_x.a(\lambda_t b(x,t,f(\lambda_u a(u,t,z),\lambda_s w)),w,y) \tag{5.1}$$

The first step is to add a global binding context for the three free variable instances z, w and y. The global context can include binders for other, unrelated variables like v below.

$$\lambda_z.\lambda_v.\lambda_w.\lambda_y.[\ ] \tag{5.2}$$

Under this assumption all variables become bound and therefore each variable occurrence can be tagged with integers d and l standing for the *reference depth* and the *binding level* of this variable occurrence. An instance $x_{d,l}$ signifies that there are d - 1 intervening abstractions between x and its binder[30] and that the total number of abstractions between x and the "root" of the global context is l. The de Bruijn (*dB*) index d refers to an originally free variable if and only if it exceeds the level l. With these conventions the example expression 5.1 is decorated as follows:

$$\lambda_x.a(\lambda_t b(x_{2,2}, t_{1,2}, f(\lambda_u a(u_{1,3}, t_{2,3}, z_{7,3}), \lambda_s w_{5,3})), w_{3,1}, y_{2,1}). \qquad (5.3)$$

Without loss of information all identifiers can be dropped yielding the "name-free" expression:

$$\Omega_{dB} \equiv \lambda.a(\lambda b(2,1, f(\lambda a(1,2,7), \lambda 5)), 3, 2). \qquad (5.4)$$

The equivalent *BTF* expression for Ω uses negative integers to denote the bindings and Curry'ed operators for the function symbols:

$$\Omega_{BTF} \equiv (-1 \ a \ (-1 \ b \ 1 \ 0 \ (f \ (-1 \ a \ 0 \ 1 \ 6) \ (-1 \ 4))) \ 2 \ 1).$$

It is easier to verify the validity of these manipulations and to check the values of the indices while using the 2-dimensional depiction of Figure 5.1.

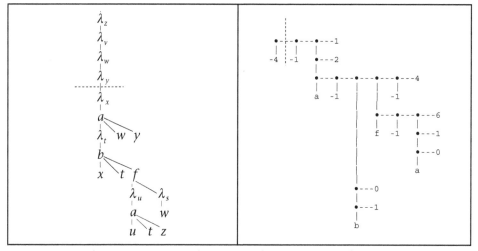

Figure 5.1 Example expression Ω in de Bruijn and "raw" BTF tree forms.

---

[30] This is the classic convention. In our treatment indices start at zero so their values are exactly the number of the intervening abstractions.

The next major step is to define the rules of substitution for name-free expressions. This step is important because it specifies how *dB* indices are to be interpreted operationally. We follow closely de Bruijn's exposition but we list only the most interesting cases. The definition will feel familiar to the reader because its rules are similar to the manipulations in *BTF* notation and the HOR processing of the previous chapter. The notation is slightly different; terms are enclosed in $\langle\rangle$ brackets and the substitution is written as $\langle S(\dots); \langle\rangle\rangle$ where ellipses denote the sequence of substitutions. Under this convention

$$\langle S(\dots,\langle \Sigma_3 \rangle,\langle \Sigma_2 \rangle,\langle \Sigma_1 \rangle; \langle \Omega \rangle)\rangle \tag{5.5}$$

denotes the result of substituting expressions $\Sigma_1 \dots \Sigma_n$ for the free variables of Ω with indices $1\dots n$. The sequence can be infinite but only finitely many Σ_k's are relevant.

We start with applications. In this case substitution distributes over the operator and operand without complications.

$$\langle \Omega \rangle = \langle \Omega_1 \rangle,\langle \Omega_2 \rangle$$
$$\langle S(\dots,\langle \Sigma_3 \rangle,\langle \Sigma_2 \rangle,\langle \Sigma_1 \rangle; \langle \Omega \rangle)\rangle = \langle \Gamma_1 \rangle,\langle \Gamma_2 \rangle \tag{5.6}$$
$$\Gamma_i = \langle S(\dots,\langle \Sigma_3 \rangle,\langle \Sigma_2 \rangle,\langle \Sigma_1 \rangle; \langle \Omega_i \rangle)\rangle$$

Next we consider the case of a variable. The result expression is obtained by direct indexing into the value sequence.

$$\langle \Omega \rangle = k$$
$$\langle S(\dots,\langle \Sigma_3 \rangle,\langle \Sigma_2 \rangle,\langle \Sigma_1 \rangle; \langle \Omega \rangle)\rangle = \Sigma_k \tag{5.7}$$

By now it should be clear that the sequence of expressions constitutes an environment. In the *BTF* notation and in the theories which follow expressions like (5.7) are not just a "meta" level construction but also valid terms. Disregarding for a moment that indexing in this case starts at one the term above is written in *BTF* as

$$[k \ \{\Sigma_1 \ \Sigma_2 \ \Sigma_3 \ \dots\}]. \tag{5.8}$$

The most interesting case is that of an abstraction. When a substitution is pushed over an abstraction two things must happen. First, variables which refer to the binder that was just crossed are given the index 1; this preserves the original binding. Second, the value of the indices corresponding to the relatively free variables in each of the elements of the environment sequence are "offset" by one so that they still reach their original binders.

$$\langle \Omega \rangle = \lambda\langle \Gamma \rangle$$
$$\langle S(\dots,\langle \Sigma_3 \rangle,\langle \Sigma_2 \rangle,\langle \Sigma_1 \rangle; \langle \Omega \rangle)\rangle = \lambda\langle S(\dots,\langle \Lambda_3 \rangle,\langle \Lambda_2 \rangle,\langle \Lambda_1 \rangle,1; \langle \Gamma \rangle)\rangle \tag{5.9}$$
$$\langle \Lambda_i \rangle = \langle S(\dots,4,3,2; \langle \Sigma_i \rangle)\rangle$$

In other words each $\langle \Lambda_i \rangle$ is obtained from $\langle \Sigma_i \rangle$ by adding one to all indices which denote relatively free variables.

Since constructions of the form $\langle S(\dots, n+2, n+1; \langle \Sigma \rangle) \rangle$ are so common, de Bruijn suggests that they be abbreviated by $\tau_n(\langle \Sigma \rangle)$. The τ_n operator alters substitutions to compensate for the abstraction context. It is equivalent to the "↑" operator of the next section ($\tau_n(\langle \Sigma \rangle) = \Sigma \uparrow^n$) and plays the same role as the ϕ count in HOR. The seeds of the *UVC* mechanism of HOR which updates on-demand the indices of relatively free variables are present in the following statement:

> ... In automatic formula manipulation it may be a good strategy to refrain from evaluating such $\tau_n \langle \Sigma \rangle$ but just to store them as pairs l, $\langle \Sigma \rangle$, and go into (full or partial) evaluation only if necessary.... [Bruijn 72], p386.

Mutual application of name-free expressions is made explicit with the notation $A(\cdot, \cdot)$. After β-reduction the expression $A(\lambda \langle \Omega \rangle, \langle \Gamma \rangle)$ is $\langle S(\dots, 3, 2, 1, \langle \Gamma \rangle; \langle \Omega \rangle) \rangle$. Conversely, η-reduction is the transformation of the expression $\lambda A\ (\langle \tau_1(\langle \Lambda \rangle) \rangle, 1) \rightarrow\!\!\!\!\!\!- \Lambda \rangle$.

Finally, composition of substitutions is captured by the following rule.

$$
\begin{aligned}
\langle \Omega \rangle &= \langle S(\dots, \langle \Lambda_2 \rangle, \langle \Lambda_1 \rangle; \langle \Lambda \rangle) \rangle \\
\langle S(\dots \langle \Sigma_2 \rangle, \langle \Sigma_1 \rangle; \langle \Omega \rangle) \rangle &= \langle S(\dots, \langle \Gamma_2 \rangle, \langle \Gamma_1 \rangle; \langle \Lambda \rangle) \rangle \\
\langle \Gamma_i \rangle &= \langle S(\dots \langle \Sigma_2 \rangle, \langle \Sigma_1 \rangle; \langle \Lambda_i \rangle) \rangle
\end{aligned}
\tag{5.10}
$$

We conclude this short presentation of de Bruijn's system by reiterating that the name-free expressions and all their manipulations always correspond to "name-carrying" ones if a context of sufficient depth for globally free variables is provided.

5.2 EXPLICIT SUBSTITUTIONS - ΛCCL

More recently Abadi, Cardelli, Curien and Lévy have put forward a series of formal systems which include substitutions as an explicit part of the calculus [Abadi 90], [Abadi 91]. This treatment makes the connection between theoretical investigations of the properties of such calculi and the practical concerns of implementations of reduction. Independently and at about the same time John Field has proposed a nearly identical theory as a tool for studying the optimality properties of various reduction strategies [Field 90], [Field 91]. The following brief sketch is based on the untyped version of the former theory but all of what we cover is also expressible in Field's ΛCCL.

The theory of explicit substitutions is embodied by an extended Lambda Calculus with two sorts of terms:

(*a*) terms of the classic calculus in a de Bruijn notation, and

(*b*) terms representing classic calculus terms subject to accumulated substitutions.

The portion of the theory which manipulates substitutions is given the name σ and the whole system is called the λσ -calculus. The authors introduce it with the following passage.

> … In some respects the λσ -calculus resembles the calculi of combinators, including those of categorical combinators [Curry 58]. The λσ -calculus and the combinator calculi all give full formal accounts of the process of computation, without suffering from unpleasant complication in the (informal) handling of variables. They all make it easy to derive machines for the λ-calculus and to show the correctness of these machines. From our perspective, the advantage of the λσ -calculus over combinator calculi is that it remains closer to the original λ-calculus. … [Abadi 91], p2.

The syntactic elements of σ are listed in Table 5.1. The syntax of the terms of the untyped λσ -calculus is as follows.

$$
\begin{aligned}
&Terms \quad a := 1 \,|\, ab \,|\, \lambda a \,|\, a[s] \\
&Substitutions \quad s := id \,|\, \uparrow \,|\, a \cdot s \,|\, s \circ t
\end{aligned}
\tag{5.11}
$$

The manipulations put forward by this calculus should sound natural to readers who have absorbed the previous discussion and are already familiar with the de Bruijn notation. Variables are denoted by positive integers starting with 1. We follow closely Abadi et al.

- id is the identity substitution {1/1, 2/2, …} which does not affect any indices. It is also written as {i/i}.

- $\uparrow$ denotes the substitution {$(i + 1)/i$}. For example, 1[$\uparrow$] = 2. With the inclusion of this operator one can see that only the index 1 is really needed; n+1 can be coded as 1[$\uparrow^n$], where $\uparrow^n$ is the composition of n shifts, $\uparrow \circ \cdots \circ \uparrow$. One can also write $\uparrow^0$ for id.

- $i[s]$ is the value of the de Bruijn index i under the substitution s. It is also written as $s(i)$.

- $a \cdot s$ is the cons of a onto s. It denotes the substitution {$a/1, s(i)/(i+1)$} which maps the term a to the variable with index 1 and the elements of s in the same order to the rest of

id	*Identity*
$i[s]$	*Value*
$\uparrow$	*Shift*
$a \cdot s$	*Cons*
$s \circ i$	*Composition*

Table 5.1 Substitution operators of the λσ-calculus.

the variables. For example,

$$a \cdot id = \{a/1, 1/2, 2/3, \ldots\}$$
$$1 \cdot \uparrow = \{~1/1, \Uparrow 1)/2, \Uparrow 2)/3, \ldots\} = id$$

- $s \circ t$ is the composition of substitutions s and t. If a is a term $a[s \circ t] = a[s][t]$, and hence, $s \circ t = \{s(i)/i\} \circ t = \{s(i)[t]/i\}$. Each element of s is affected by the substitution t before the composition can be applied. Some laws of composition are

$$id \circ t = \{id(i)[t]/i\} = \{t(i)/i = t$$
$$\uparrow \circ (a \cdot s) = \{\Uparrow i)[a \cdot s]/i\} = \{s(i)/i\} = s.$$

The second identity warrants further illumination; Why is $\uparrow(i)[a \cdot s]$ equal to $s(i)$? Consider the infinite substitution $\Uparrow i)$. It takes a de Bruijn index i and turns it into $i + 1$. Then the cons $[a \cdot s]$ turns the non-existent index 1 to a and the index $i + 1$ to $s(i)$. Therefore the result of $\uparrow(i)[a \cdot s]$ is identical to that of substitution s. Via this example we see that $\uparrow \circ s$ is the *rest* of s, i.e., s without its first component. Formally, $1[s] \langle \uparrow \circ s) = s$.

- Substitution distributes over application without any complications.

$$(ab)[s] \rightarrow (a[s])(b[s])$$

- An index of 1 modified by a substitution which has c as its first element is equivalent to a variable look-up.

$$1[c \cdot s] \rightarrow c$$

- Again, the most interesting law is the one which governs how a substitution distributes over an abstraction. The proper substitution is one which intercepts the relatively free occurrences referring to the binder λ and has all its elements "shifted" by one since they now have to operate under an abstraction context augmented by one binder.

$$(\lambda a)[s] \rightarrow \lambda(a[1 \cdot (s \circ \uparrow)])$$

- Finally, the composition of two substitutions the first of which has a term a as its first element is written as

$$(a \cdot s) \circ t = a[t] \cdot (s \circ t).$$

Table 5.2 lists the axioms of the calculus together with four rules that are added to augment confluence properties.

Beta	$(\lambda a)b = a[b \cdot id]$
Clos	$a[s][t] = a[s \circ t]$
IdL	$id \circ s = s$
ShiftId	$\uparrow \circ id = \uparrow$
ShiftCons	$\uparrow \circ (a \cdot s) = s$
VarId	$1[id] = 1$
VarCons	$1[a \cdot s] = a$
Abs	$(\lambda a)[s] = \lambda(a[1 \cdot (s \circ \uparrow)])$
Map	$(a \cdot s) \circ t = a[t] \cdot (s \circ t)$
App	$(ab)[s] = (a[s])(b[s])$
Ass	$(s_1 \circ s_2) \circ s_3 = s_1 \circ (s_2 \circ s_3)$
— —	
Id	$a[id] = a$
IdR	$s \circ id = s$
VarShift	$1 \cdot \uparrow = id$
Scons	$1[s] \cdot (\uparrow \circ s) = s$

Table 5.2 (Abadi et al) The axioms of the $\lambda\sigma$-calculus.

When terms of the $\lambda\sigma$ -calculus are examined one must be careful to interpret correctly the binding indices. In the term

$$(\lambda(1[2 \cdot id]))[a \cdot id]$$

the occurrence of the variable with index **1** is not bound by the surrounding abstraction. The substitution $[2 \cdot id]$ takes precedence; the correct binding is the subterm $[a \cdot id]$ which gives the value a to the overall term.

The same expression in *BTF* reveals that there is a one-to-one correspondence between the two notations; indices are offset by one, the abstraction is represented by a negative integer and the delayed substitutions are written in the form [*term* {*envlist*}].

$$[(-1[0\{[1\{\}]\}])\{[a\{\}]\}]$$

In order to make reduction efficient Abadi et al propose that a new rule (5.12), is included which is justified under σ and *IdR*. It has precedence over the *Beta* and *Abs* rules and corresponds to pushing a suspension onto the environment.

$$((\lambda a)[s])b \overset{wn}{\to} a[b \cdot s] \tag{5.12}$$

A weak strategy $\overset{wn}{\to}$, follows directly from the previous rule and the axioms of Table 5.2. This weak strategy can be readily adapted to strong reduction because the explicit substitution operators make it possible to specify the needed transformations while operating in the scope of lambdas. The extension to *nf* is captured by the following two rules:

$$\frac{wn(a,s) = (\lambda a', t)}{nf(a,s) = \lambda(nf(a',1 \cdot (t \circ \uparrow)))} \tag{5.13}$$

$$\frac{wn(a,s) = (\mathbf{n}(a_1[s_1]) \ \ldots \ (a_m[s_m]), \ id)}{nf(a,s) = (\mathbf{n}(nf(a_1,s_1) \ \ldots \ nf(a_m,s_m))} \tag{5.14}$$

Terms of the form $(\lambda a)[s]$ correspond to our operational *weak* normal forms which need to be unraveled for strong normalization. Terms of the form $\mathbf{n}a_1 \ldots a_n$ correspond to our operational *head* normal forms with the top-level abstraction context removed.

Once the non-determinism of the rules is removed a *whnf* reducer takes the form shown in the upper portion of Table 5.3. The machine is started with the triplet $\langle s, a, \{\} \rangle$. The rules of the last two rows implement the recursive invocations required by the inference rules (5.13) and (5.14).The recursive call for the operand during the evaluation of an application is delayed by using a stack to record the operand and the applicable substitution.

When the machine stops with a term which is a top-level abstraction

$$\langle s, \lambda a, \{\} \rangle$$

it is restarted with the body of the abstraction, a null stack and properly modified substitution:

$$\langle \mathbf{1} \cdot (s \circ \uparrow), a, \{\} \rangle$$

When the machine stops with an "ultimately" processed head variable and a list of substitutions in the stack

$$\langle id, \mathbf{n}, a_1[s_1] \mathbin{:} \ldots a_m[s_m] \cdot \{\} \rangle$$

m instances of the reducer are started with each of the substitutions and an empty stack:

$$\langle s_1, a_1, \{\} \rangle, \ldots \langle s_n, a_n, \{\} \rangle$$

For purposes of comparison, we repeat in Table 5.4 the abstract machine for pure Head Order Reduction. We remind the reader that de Bruijn variables in the case of HOR start at zero. The advantages of HOR are the streamlining of the processing of unbound variables via the ϕ counts and the implementation of the recursive processing for subterms emanating from the *hnf* via the lambda and join markers.

Subst	Term	Stack	Subst	Term	Stack
$\uparrow$	**n**	S	id	**n** $+1$	S
$a[s]\cdot t$	**1**	S	s	a	S
$a\cdot s$	**n** $+1$	S	s	**n**	S
$s\circ s'$	**n**	S	s'	**n**$[s]$	S
s	ba	S	s	b	$a[s]\cdot S$
s	λa	$b[t]\cdot S$	$b[t]\cdot s$	a	S
s	**n**$[id]$	S	s	**n**	S
s	**n**$[\Uparrow]$	S	s	**n** $+1$	S
s'	**1**$[a\cdot s]$	S	s'	a	S
s'	**n** $+\mathbf{1}[a\cdot s]$	S	s'	**n**$[s]$	S
s''	**n**$[s\circ s']$	S	$s'\circ s''$	**n**$[s]$	S
s'	$a[s]$	S	$s\circ s'$	a	S
s	λa	$\{\}$	$\lambda\left(1\cdot(s\circ\uparrow)\right)$	a	$\{\}$
id	**n**	$a_1[s_1]{:}\cdots\cdot a_n[s_n]\cdot\{\}$	$\mathbf{n}\left(\begin{array}{c}s_1\\ \ldots\\ s_n\end{array}\right.$	$\begin{array}{c}a_1\\ \ldots\\ a_n\end{array}$	$\left.\begin{array}{c}\{\}\\ \ldots\\ \{\}\end{array}\right)$

Table 5.3 (Abadi et al) Untyped $\lambda\sigma$-calculus $\overset{wn}{\to}$ abstract machine.

Term	Stack	Env	UVC	Dir	Term	Stack	Env	UVC	Dir
ab	s	ρ	ϕ	$\downarrow$	a	$b\{\rho\}{:}s$	ρ	ϕ	$\downarrow$
λa	$c{:}s$	ρ	ϕ	$\downarrow$	a	s	$c\bullet\rho$	ϕ	$\downarrow$
λa	s	ρ	ϕ	$\downarrow$	a	$\lambda{:}s$	$\phi-1\bullet\rho$	$\phi-1$	$\downarrow$
$\mathbf{i}+1$	s	$u\bullet\rho$	ϕ	$\downarrow$	$\mathbf{i}$	s	ρ	ϕ	$\downarrow$
0	s	$a\{\rho'\}\bullet\rho$	ϕ	$\downarrow$	a	s	ρ'	ϕ	$\downarrow$
0	s	$\phi'\bullet\rho$	ϕ	$\downarrow$	$\phi'-\phi$	s	—	ϕ	$\uparrow$
a	$\lambda{:}s$	—	ϕ	$\uparrow$	λa	s	—	$\phi+1$	$\uparrow$
b	$@{:}a{:}s$	—	ϕ	$\uparrow$	ab	s	—	ϕ	$\uparrow$
a	$b\{\rho'\}{:}s$	—	ϕ	$\uparrow$	b	$@{:}a{:}s$	ρ'	ϕ	$\downarrow$

Table 5.4 Head Order Reduction abstract machine.

As mentioned in the beginning of this section, Field's ΛCCL is captured by a similar set of axioms and obeys an equivalent set of laws (Table 5.5). A normalizing procedure *rwhnf()* is described in terms of the structure of a ΛCCL term and in terms of a procedure *rpenf()*, which transforms environments in a form (*partial environment normal form, PENF*) in which "shift" operators are fully effected.

$$
\begin{array}{ll}
Beta & \mathrm{Apply}\,(\Lambda(A), B) = [A, \langle \varnothing, B \rangle] \\[4pt]
AssC & [[A, E_1], E_2] = [A, E_1 \circ E_2] \\[4pt]
NullEL & \varnothing \circ E = E \\[4pt]
NullER & E \circ \varnothing = E \\[4pt]
ShiftE & \uparrow \circ \langle E, A \rangle = E \\[4pt]
VarRef & [\mathrm{Var}, \langle E, A \rangle] = A \\[4pt]
D\Lambda & [\Lambda(A), E] = \Lambda([A, \langle E \circ \uparrow, \mathrm{Var} \rangle]) \\[4pt]
DE & \langle E_1, A \rangle \circ E_2 = \langle E_1 \circ E_2, [A, E_2] \rangle \\[4pt]
DApply & \mathrm{Apply}\,((A, B), E) = \mathrm{Apply}\,([A, E], [B, E]) \\[4pt]
AssE & (E_1 \circ E_2) \circ E_3 = E_1 \circ (E_2 \circ E_3) \\[4pt]
NullC & [A, \varnothing] = A
\end{array}
$$

Table 5.5 (Field) The axioms of ΛCCL.

5.3 EXAMPLE AND COMPARISON WITH HOR

We conclude this short chapter by showing a concrete example of reduction in all three systems and traces via *HOR* and *AP*. This exercise reinforces the similarities of these systems and makes it clear that an implementation can rely on operational elements that achieve directness while remaining faithful to the underlying theory.

The term $M \equiv \lambda x.(\lambda y.\,y)\,x$ has the following ΛCCL reduction when one is allowed to process the inner redex first. Encoding of the two bound variables into de Bruijn indices yields the term $\Lambda(Apply(\Lambda(Var), Var))$. Two reduction steps suffice:

$$
\Lambda(Apply(\Lambda(Var), Var)) \to_{(Beta)}
$$
$$
\Lambda[Var, \langle 0, Var \rangle]) \to_{(VarRef)} \Lambda(Var)
$$

The term M is an abstraction with an inner β-redex. In the de Bruijn notation of Abadi et al it is written as $(\lambda(\lambda 1)1)$. In order to become a legal focus for the $\overset{wn}{\to}$ algorithm it must be packaged as the pseudo-closure $(\lambda(\lambda 1)1)[id]$. The reduction steps in the $\lambda\sigma$-calculus are as follows:

$$
\begin{aligned}
&(\lambda(\lambda 1)1)[id] \to \\
&(\lambda(((\lambda 1)1)[1 \cdot (id \circ \uparrow)])) \to \\
&(\lambda(((\lambda 1)[1 \cdot (id \circ \uparrow)])(1[1 \cdot (id \circ \uparrow)]))) \to \\
&(\lambda(1[(1[1 \cdot (id \circ \uparrow)]) \cdot 1 \cdot (id \circ \uparrow)])) \to \\
&(\lambda(1[1 \cdot (id \circ \uparrow)])) \to \\
&(\lambda 1)
\end{aligned}
$$

Next, we show the reduction of M via the non-deterministic rules of the HOR machinery of section 3.6. Under our convention the term is written as $(\lambda(\lambda 0)0)$. The reduction sequence starting with the rule for beta redices is listed below. For quick reference Table 5.6 repeats the non-deterministic reduction rules of Head Order Reduction which transform term, environment and abstraction depth triplets.

$$(\lambda(\lambda 0)0) \ \{\} \ \ 0 \rightarrow$$
$$(\lambda 0) \ \{0\} \ \ 0 \rightarrow$$
$$\lambda(0 \ \{-1 \ \ 0\} \ -1) \rightarrow$$
$$\lambda 0$$

$((\lambda a)b) \ \rho \ \phi$	$\rightarrow$	$a \ b\{\rho\} \cdot \rho \ \ \phi$	
$(ab) \ \rho \ \phi$	$\rightarrow$	$(a \ \rho \ \phi)(b \ \rho \ \phi)$	
$(\lambda a) \ \rho \ \phi$	$\rightarrow$	$\lambda(a \ \phi - 1 \cdot \rho \ \ \phi - 1)$	
$i \ \ c \cdot \rho \ \phi$	$\rightarrow$	$i - 1 \ \rho \ \phi$	
$0 \ a\{\rho'\} \cdot \rho \ \phi$	$\rightarrow$	$a \ \rho' \ \phi$	
$0 \ \ \phi' \cdot \rho \ \phi$	$\rightarrow$	$\phi' - \phi$	

Table 5.6 Non-deterministic reduction rules for Head Order Reduction.

Finally, in the machine-oriented *BTF* notation term *M* is written (-1 (-1 0) 0). Tables 5.7 and 5.8 show the *HOR* and *AP* reduction traces. A comparison to the previous sequences shows that these reducers avoid the formation of substitution redices (*a*) by delaying their effect, (*b*) by providing "random access" into the environment structure and (*c*) in the case of HOR, by utilizing the scalar ϕ count to correct on-demand the indices of relatively free variable occurrences.

step	term	stack	env	phi
D0	(-1 (-1 0) 0)	nil	nil	0
D1	((-1 0) 0)	(\)	(-1)	-1
D2	(-1 0)	(-1 \)	(-1)	-1
•D3	0	(\)	(-1 -1)	-1
U4	0	(\)		-1
U5	(-1 0)	nil		0

Table 5.7 HOR reduction of (-11 (-1 0) 0).

step	term/clos	env
N0	(-1 (-1 0) 0)	nil
N1	((-1 0) 0)	(g1)
L2	(-1 0)	(g1)
R2	([(-1 0) ({ g1 })])	
•N3	0	(g1 g1)
R3	g1	
R1	g1	
R0	(-1 0)	

Table 5.8 AP reduction of (-11 (-1 0) 0).

De Bruijn presented the rules of how β-reduction can be effected via literal substitution of name-free expressions. The technique proves even more useful when substitutions are delayed. The convergence of these three theories and the strong relation to the minimal reduction machinery of HOR is another indication that these ideas successfully embody the key aspects of Lambda Calculus formula manipulation.

Chapter 6

Sharing Strategies

6.1 GENERAL

The major concern of evaluation of lambda expressions is to avoid unnecessary effort by choosing to contract redices in a sequence which avoids duplicated work. Optimality is defined as the minimization of the length of a reduction sequence. Lévy has developed the framework of "parallel reductions" in which this question can be studied formally.

> ... This short notice is to refresh an old problem that several researchers have tried to tackle unsuccessfully. This problem originated by Wadsworth's Ph.D. dissertation can be fortunately easily stated: how to evaluate efficiently lambda expressions? By efficient, we mean optimal which may not be the same. And optimal means without duplication of contraction of redices. ... [Lévy 88].

But optimality has proved an elusive goal. Not only is it difficult to achieve, but all indications are that to even approach it requires efforts that wipe out any potential gains from the avoidance of reductions. Structure sharing via graph representations is the usual pragmatic recourse.

Several mechanisms have been suggested to edge closer towards the ideal of optimality. The most promising studies rely on delayed substitutions as part of the overall strategy.

> ... The question then arises as to whether some combination of shared environments, closures and terms could be uses to achieve an optimal reduction scheme or at least improve on Wadsworth's method. ...[Field 90], p3.

Barendregt et al. [Barendregt 86] propose to shift the focus from the leftmost redex to the redices that are actually *needed* for a normal form to be reached. They develop a theory of reduction based on whether the reduction of a subterm is needed. One of their conclusions is that the head normal form and the left spine is the natural half-way stage by which "neededness", while generally undecidable, may be effectively determined.

> ... Just which of the desired algorithms is appropriate for a given implementation is technology and application dependent. Our contribution is to offer a range of choices to the implementor which frees him from the sharp distinction between applicative and normal order strategies, which currently forces him to either accept wholesale the inefficiency risks associated with normal order, or to buy the known efficiency of applicative order at the cost of losing normalising properties for his implementation.... [Barendregt 86].

But the selection of a redex cannot rely on a complicated decision procedure. It has to be direct and rely on simple control information so that it is fast when mechanized. Accomplishing a reduction, literally or in a delayed manner, does not have large enough "value" to justify significant overhead. Sharing and its overhead pays off only when it costs less than the savings it can provide.

Empirical evidence gained from many challenging expressions suggests that the efficiency associated with applicative order and weak normalization can indeed be matched— even exceeded—by normal order strategies which incorporate certain degree of sharing (cf. sections 7.2 and 7.3).

A factor which frequently appears to be overlooked is that a large part of the unnecessary costs incurred during normal-order reduction without sharing is not due to the recomputation of the normal forms of expressions strictly in argument position (*passive* occurrences) but of those in head (operator or *active*) position. The nature of most problems is such that during the application of a strongly normalizing algorithm, argument expressions do get referenced from head variables and hence benefit primarily by a strategy that avoids extraneous evaluation by recording partially reduced expressions that appear in operator position.

The choice that leads to the concrete *HOR-SH* is the one which allows us to be as conservative with respect to reduction counts as possible; viewed in a complementary manner to be as eager with respect to normalizing as possible. The suspension referred to by a head variable is strongly reduced in isolation. Then its normal form is allowed to act as an operator with the actual arguments present. Sharing is accomplished through the environment; the suspension entry is updated to reflect the result of the reduction in isolation before the reduced term is reentered with the original spine restored.

The introduction of sharing destroys the locality of β-contractions. Without sharing every contraction takes the expression one step closer to *nf* (if one exists) along a certain path which is defined by the overall strategy. This is true for both literal and delayed substitution algorithms. In fact, even under delayed substitutions there is at each step a mapping to an equivalent fully-unraveled expression in which the substitutions are effected literally. With sharing it is no longer possible to devise such a simple correspondence. Many, seemingly unrelated portions of the unraveled expression may be affected in each step.

In this chapter we present an augmentation of the basic Head Order Reducer which achieves a significant degree of sharing. It is designed to fit well with the natural operational aspects of *HOR*. Before we present it in detail, we discuss briefly the augmentation with sharing of the *RTNF/RTLF* reducer which we take as the reference point because it provides a degree of sharing similar to the one present in Wadsworth reduction.

6.2 SHARING IN RTNF/RTLF REDUCERS

Aiello–Prini have showed how one can augment the delayed *RTNF/RTLF* reducer with sharing. When a suspension that is referred to by a variable is reduced, the result is allowed to overwrite the environment entry that holds the suspension. The mode of this reduction may be either weak or strong; therefore the result is either a normal form or a closure.

There are two fine points that one must heed. After updates of environment entries have occurred, a variable lookup can dereference to an abstraction. In case of *rtlf* a closure is always returned so the abstraction is packaged as a closure with a null environment. The other complication is that in case a variable cannot be located in the environment then the proper value to be returned is a pointer to the variable itself.

This strategy is ingenious because it manages to exhibit a great degree of savings for expressions that include opportunities for sharing without the overhead of additional reductions in isolation. This fact is reflected in the performance measurements of chapter 7. The *RTNF/RTLF* strategy, sharing included, can be cast with HOR-style machinery with an additional gain in performance.

We proceed to illustrate now the workings of the *AP-SH* reducer with the addition of sharing. As an example we choose the expression (-2 (-1 0 (0 2)) ((-2 0 1) 0)) which is a simplified variant of the one chosen by Wadsworth [Wadsworth 71], p.172.

In step *N4* the focus is the combination (0 (0 2)). Its operator is dereferenced in step *L6* and after the *rtlf* reduction of the suspension [((-2 0 1) 0) { g2 g1 }] is completed, the resulting closure (step *R5)* overwrites the original environment entry (shown underlined).

step	term/susp	env
N0	(-2 (-1 0 (0 2)) ((-2 0 1) 0))	nil
N1	(-1 (-1 0 (0 2)) ((-2 0 1) 0))	(g1)
N2	((-1 0 (0 2)) ((-2 0 1) 0))	(g2 g1)
L3	(-1 0 (0 2))	(g2 g1)
R3	[(-1 0 (0 2)) { g2 g1 }]	
•N4	(0 (0 2))	([((-2 0 1) 0) { g2 g1 }] g2 g1)
L5	0	([[((-2 0 1) 0) { g2 g1 }] g2 g1)
L6	((-2 0 1) 0)	(g2 g1)
L7	(-2 0 1)	(g2 g1)
R7	[(-2 0 1) { g2 g1 }]	
•L8	(-1 0 1)	([0 { g2 g1 }] g2 g1)
R8	[(-1 0 1) { [0 { g2 g1 }] g2 g1 }]	
R6	[(-1 0 1) { [0 { g2 g1 }] g2 g1 }]	
R5	[(-1 0 1) { [0 { g2 g1 }] g2 g1 }]	
•N9	(0 1)	([(0 2) { [(-1 0 1) { [0 { g2 g1 }] g2 g1 }] g2 g1 }] [0 { g2 g1 }] g2 g1)
L10	0	([(0 2) { [(-1 0 1) { [0 { g2 g1 }] g2 g1 }] g2 g1 }] [0 { g2 g1 }] g2 g1)
L11	(0 2)	([(-1 0 1) { [0 { g2 g1 }] g2 g1 }] g2 g1)
L12	0	([[(-1 0 1) { [0 { g2 g1 }] g2 g1 }] g2 g1)
L13	(-1 0 1)	([0 { g2 g1 }] g2 g1)
R12	[(-1 0 1) { [0 { g2 g1 }] g2 g1 }]	
•L14	(0 1)	([2 { [(-1 0 1) { [0 { g2 g1 }] g2 g1 }] g2 g1 }] [0 { g2 g1 }] g2 g1)
L15	0	([2 { [(-1 0 1) { [0 { g2 g1 }] g2 g1 }] g2 g1 }] [0 { g2 g1 }] g2 g1)
L16	2	([(-1 0 1) { [0 { g2 g1 }] g2 g1 }] g2 g1)
R16	g1	
R15	g1	
N17	1	(g1 [0 { g2 g1 }] g2 g1)
N18	0	(g2 g1)
R18	g2	
R17	g2	
R10	(g1 g2)	
N19	1	((g1 g2) g2 g2 g1)
R19	g2	
R2	(g1 g2 g2)	
R1	(-1 g1 0 0)	
R0	(-2 1 0 0)	

Table 6.1 AP-SH reduction of wads45 = (-2 (-1 0 (0 2))((-2 0 1) 0)) .

•N4:	(0 (0 2))	(([((-2 0 1) 0) ({ g2 g1 })]) g2 g1)
L5:	0	(([((-2 0 1) 0) ({ g2 g1 })]) g2 g1)
L6:	((-2 0 1) 0)	(g2 g1)

In step *N9* the term portion of the result closure participates in a reduction which pushes the pending subterm (0 2) of step *N4* onto the environment. When this term is referenced at step *L11* the variable which points to the topmost environment entry finds the weakly reduced subterm and extra work is avoided.

R5: ([(-1 0 1) ({ ([0 ({ g2 g1 })]) g2 g1 })])
•*N9:* (0 1)   (([(0 2) ({ ([(-1 0 1) ({ ([0 ({ g2 g1 })]) g2 g1 })]) g2 g1 })]) ([0 ({ g2 g1 })]) g2 g1)

L11: (0 2) $(([\ (\text{-}1\ 0\ 1)\ ([\ ([\ 0\ ([\ g2\ g1\ ])\ ])\ ]\ g2\ g1\ ])\ ])\ g2\ g1)$
L12: 0 $(([\ (\text{-}1\ 0\ 1)\ ([\ ([\ 0\ ([\ g2\ g1\ ])\ ])\ ]\ g2\ g1\ ])\ ])\ g2\ g1)$

The complete trace is listed in Table 6.1. Plain *AP* reduction of the same expression can be found in Appendix B in Table B.5.

6.3 HEAD ORDER REDUCTION WITH SHARING

The previous discussion of mechanisms for sharing supports the tactic of treating specially a subterm the first time it is referred by a variable during the course of reduction. Depending on the overall strength of the algorithm, this special processing can be one that results in weak, head or strong normal forms.

In the realm of HOR, a sharing reduction strategy is made possible by the inclusion of the abstraction-context tagging of a *strongly* reduced, but potentially out of context, expression. Berkling has introduced this tagging which is necessary when de Bruijn indices are employed because it permits a shared normal form to be interpreted correctly within contexts of nested abstractions of different degree.

Sharing under Weak HOR

We start by discussing how the weak HOR can be augmented with sharing. When a suspension is referenced from a variable in head position the suspension is (weakly) reduced in isolation—i.e., with a null stack. The environment is updated to point to the result which is constrained to be a pure closure. Reduction continues with this closure and the restored stack. This sharing tactic turns out to be equivalent to the one provided by the *rtlf* function of Aiello–Prini (*RTLF-SH*). An abstract machine is not given since the next subsection details a superset of these actions for strong *HOR*.

Table 6.2 shows the reduction of ((-2 1 1) (-2 1 0) (-1 0)) via the modified algorithm. The reduction of the suspension [(-2 1 0) { }] in step *D6* is started with a null stack. In this case the term of the suspension is already in *whnf* and so it is simply returned as a closure. Reduction continues with this closure and the restored stack. In this example sharing provides no benefit at all. We include it to emphasize that this is quite frequently the case.

step	term/clos	stack	env
D0	((-2 1 1) (-2 1 0) (-1 0))	nil	nil
D1	((-2 1 1) (-2 1 0))	(([(-1 0) ({})]))	nil
D2	(-2 1 1)	(([(-2 1 0) ({})]) ([(-1 0) ({})]))	nil
•D3	(-1 1 1)	(([(-1 0) ({})]))	(([(-2 1 0) ({})]))
•D4	(1 1)	nil	(([(-1 0) ({})]) ([(-2 1 0) ({})]))
D5	1	(([(-2 1 0) ({})]))	(([(-1 0) ({})]) ([(-2 1 0) ({})]))
·D6	(-2 1 0)	nil	nil
R6	([(-2 1 0) ({})])	.	
D7	(-2 1 0)	(([(-2 1 0) ({})]))	nil
•D8	(-1 1 0)	nil	(([(-2 1 0) ({})]))

Table 6.2 W-HOR-SH reduction of ((-2 1 1)(-2 1 0)(-1 0)).

The next example involves an expression which does exhibit an appreciable benefit from the inclusion of sharing. In step $D3$ of Table 6.3 the variable refers to a subterm which requires three reductions to reach normal form. The subterm is reduced in isolation in steps $D4$-$D13$ and the result, (-1 0) the identity operator, is applied to itself in step $D14$. The overwriting of the environment in step $R4$ makes it possible to avoid duplicate work. In contrast, as shown in Table B.3, the pure version of weak HOR ends up reducing the subterm ((-2 1 1) (-1 0) (-1 0)) twice.

A concrete W-HOR-SH reducer is listed in Figure A.8.

step	term/clos	stack	env
D0	((-1 0 0) ((-2 1 1) (-1 0) (-1 0)))	nil	nil
D1	(-1 0 0)	([((-2 1 1) (-1 0) (-1 0)) {}])	nil
•D2	(0 0)	nil	([((-2 1 1) (-1 0) (-1 0)) {}])
D3	0	([((-2 1 1) (-1 0) (-1 0)) {}])	([((-2 1 1) (-1 0) (-1 0)) {}])
·D4	((-2 1 1) (-1 0) (-1 0))	nil	nil
D5	((-2 1 1) (-1 0))	([(-1 0) {}])	nil
D6	(-2 1 1)	([(-1 0) {}] [(-1 0) {}])	nil
•D7	(-1 1 1)	([(-1 0) {}])	([(-1 0) {}])
•D8	(1 1)	nil	([(-1 0) {}] [(-1 0) {}])
D9	1	([(-1 0) {}])	([(-1 0) {}] [(-1 0) {}])
·D10	(-1 0)	nil	nil
R10	[(-1 0) {}]	.	
D11	(-1 0)	([(-1 0) {}])	nil
•D12	0	nil	([(-1 0) {}])
D13	(-1 0)	nil	nil
R4	[(-1 0) {}]	.	
D14	(-1 0)	([(-1 0) {}])	nil
•D15	0	nil	([(-1 0) {}])
D16	(-1 0)	nil	nil

Table 6.3 W-HOR-SH-VA reduction of ((-1 0 0)((-2 1 1)(-1 0)(-1 0))).

Sharing under Strong HOR

Central in the ability to save work is the idea that the reduction of a subterm averts, in some manner, the reduction of its other instances created by previous or future contractions.

> ... The possibility of contracting redices in the common term in advance of the reduction of the entire term is limited by the fact that knowing which of the redexes in the common term will be subsequently needed is undecidable. At best, the common term can only be reduced to some "safe" form (such as *hnf*). However, this would not necessarily capture all the contractions performed in the common term during the reduction of one of its substitution instances. ... We are thus led to investigate techniques that allow the preservation of the result of contractions in the common term as a *side-effect* of reducing one of the members of a set of similar terms. ... [Field 91], p32.

There is an important difference as to the natural place for introducing sharing in the *HOR* and *AP* classes of reducers. Under the *RTNF/RTLF* strategy a suspension referred to by a variable is reduced in isolation as part of the normal procedure for reducing applications. This procedure specifies the weak reduction of the operator portion of an application so that a binding can be effected in case the weak reduction returns a closure.

Under a *HOR* strategy, contractions happen strictly along the left spine of each subterm. Reduction of a suspension referred to by a head variable continues the overall reduction of the whole expression until head normal form is reached. This implies that a *HOR down* traversal cannot return an intermediate normal form if it is started *in situ*; i.e., with the state of the machine intact, more specifically with the whole stack. There are no recursive calls to the reduction strategy where a surgical update may be inserted. It is only by omitting portions of the reversed spine that reduction can be forced to be partial.

Berkling has devised the crucial mechanism which allows the surgical update to happen in abstraction contexts of varying depth. We present next a formalization of this additional processing.

Abstract Machine for HOR with Sharing

Table 6.4 shows the additions required to the HOR abstract machine that achieve sharing. This machine builds naturally on the basis of the pure one of Table 3.8. Under sharing a new entity is introduced; a reduced subterm which may be structurally shared during the reduction process in the sense of having one instance which may be pointed to by other elements. In order for a construction of this sort to be correct—given the usual HOR representation choices—the reduced subterm needs to be annotated with the abstraction context which was in effect when it was reduced. This is so because the relatively free variables of the subterm—i.e., those who are reaching "outside" of it—have de Bruijn indices that depend on the subterm's context. We call such potentially shared subterms *shared expressions* (*shexp*s) and we refer to them by the construction $\phi.a$.

First, we describe how *shexp*s are created. When a suspension is retrieved from the environment it is strongly reduced without its applicative context. When this reduction is complete, the *nf* result is allowed to overwrite the environment location from which it was retrieved. But reduction of the overall suspension is not finished. The *nf* must be allowed to interact with any pending arguments. The possible elements of the environment are extended as shown in (6.1).

$$\begin{aligned} ShExps \ \ r &:= \ \phi.a \\ Environment \ \ \rho &:= \ \varnothing \,\big|\, c \bullet \rho \,\big|\, \phi \bullet \rho \,\big|\, r \bullet \rho \end{aligned} \qquad (6.1)$$

Now, it should be clear that this augmentation of the *HOR* abstract machine implies that a lookup can result in a the retrieval of a *shexp*. If this is the case the *shexp* does not need to be reduced further—it only needs to interact with any pending arguments. In case the *shexp* was created in an abstraction context which is identical to the one in effect and if, in addition, there are no arguments pending, the *shexp* can be incorporated as is.

Term	Stack	Env	UVC	Dir	Term	Stack	Env	UVC	Dir
0	s	$\underline{a\{\rho'\}} \bullet \rho$	ϕ	$\downarrow$	a	$\varnothing$	ρ'	ϕ	$\downarrow$
					b	s	$\varepsilon(\phi)$	ϕ	$\downarrow$
0	s	$\phi'.b \bullet \rho$	ϕ	$\downarrow$	b	s	$\varepsilon(\phi')$	ϕ	$\downarrow$

Table 6.4 Rules which add sharing in the HOR Abstract Machine.

Observe that the modified rule for a retrieved suspension has two parts. The topmost one defines the reduction in isolation, i.e., with a stack that is empty. The second part shows the subsequent reduction of the *nf* result in a special environment $\varepsilon(\phi)$ and with the original stack. Term b is the result of a recursive invocation of the overall procedure as shown in (6.2). We will later visit some optimizations which can modify the application of the second rule.

$$b = hor \ \ a \ \ \varnothing \ \ \rho' \ \ \phi \ \downarrow \qquad (6.2)$$

The in-situ update of the environment may be written as in (6.3). The underlined suspension is the one referenced via the environment lookup on the left side of the transition rule.

$$location(\underline{a\{\rho'\}}) := \ \phi.b \qquad (6.3)$$

In order to show why these decisions are sound we will have to think carefully about the meaning of an contiguous sequence of *UVC* entries. First we revisit the role of the *UVC*s.

Role of the "Unbound Variable Counts"

Consider a term H in head normal form (6.4). Suppose that among the operand expressions a suspension E_ρ is present whose term is a single variable, v_i and that this variable is embedded in an environment ρ (6.5). Now if this environment consists of entries $U_1...U_q$ then, according to the meaning of an environment described in section 3.6, v_i refers to U_i. Suppose that U_i is a *UVC* entry and that its value is the integer $-m$. Then the intended instantiation of v_i is defined to be the binder of the mth enclosing abstraction, x_m.

$$H = \lambda x_1...x_m...x_n \; . \; \upsilon \; e_1...E_\rho...e_k \tag{6.4}$$

$$E_\rho = ((\lambda v_1...v_i...v_q \; . \; v_i) \; U_1...U_i...U_q) \tag{6.5}$$

$$U_i = -m \;\; \Leftrightarrow \;\; v_i \sim x_m \tag{6.6}$$

So the role of a *UVC* entry in the environment is that of a symbolic link between a relatively free variable and a distant binder. Using the native mechanism for association, i.e., β-reduction, this construction can be written as

$$\lambda x_1...x_m...x_n \; . \; \upsilon \; e_1 \; ... \; \underline{((\lambda v_1...v_i...v_q \; . \; v_i) \; U_1...x_m...U_q)} \; ... \; e_k. \tag{6.7}$$

The pleasant attributes of *UVC*s are that (*a*) this link is clearly explained in terms of constructions native to the calculus and (*b*) only integer values are needed instead of pointers at the level of an implementation.

Meaning of a "Dummy" Environment

The special environment $\varepsilon(\phi)$ is now explained. The partial result $\phi.b$ is in *nf* so at first glance it may seem that an environment is superfluous. The fact though that the reduction in isolation is materialized under a non-null abstraction context means that the resulting *nf* will typically have relatively free variables whose indices need adjusting. Instead of literally performing such an adjusting, the meaning of these variables is maintained if the proper environment is established. A suitable environment consists of an ascending sequence of *UVC* values starting with the abstraction context ϕ.

$$\varepsilon(\phi) = \{\phi \quad \phi+1 \; ... \; -2 \; -1\} \tag{6.8}$$

Armed with the formalization of the meaning of *UVC* entries we can return to the meaning of the *BTF* expression of (6.8). Given that the environment consists solely of *UVC*s the environment is the context that is provided to a term which may occupy the square brackets in (6.9).

$$\varepsilon(\phi) = ((\lambda x_{-1}...x_\phi . [\ ]) -1 \ ... \ \phi)$$
$$= \lambda x_{-1}...x_\phi . \upsilon \ ... \ \underline{((\lambda x_{-1}...x_\phi . [\ ]) \ x_{-1} \ ... \ x_\phi)} \ ... \qquad (6.9)$$

The environment $\varepsilon(\phi)$ achieves the artificial closing of a term which is in normal form but which may have variable indices that are reaching "outside" the term. The consequence is that such terms can be used safely as if they were closed terms.

Summarizing, the end-result of the traversal of the reduced term in the ϕ-specified dummy environment—with or without an applicative context—is an on-demand update operation which produces finally correct de Bruijn indices for variables that remain relatively free in the new context.

We note at this point that in a competent implementation the dummy environment does not have to be explicitly created. One should use a special entry on the trailing portions of existing environment structures to accomplish the same effect via the look-up procedure.

The procedural interpretation of this insight is that the *hor* machine itself can be used to correct the indices of an "out-of-context" expression. Examples of the result of such an invocation on open *nf*'s are shown in Table 6.5.

s	ϕ	ϕ'	$t = hor \ s \ \varnothing \ \varepsilon(\phi) \ \phi' \downarrow$
(-1 0 3)	-3	-2	(-1 0 2)
(-1 0 2)	-3	-6	(-1 0 5)
(-1 0 2 (-1 1 3)))	-2	-5	(-1 0 (5 (-1 1 6)))

Table 6.5 Using the HOR machinery for correcting artificially closed terms.

This potential sharing of reduced but open subterms can benefit from an illustration. Figure 6.1 shows the correction that must be applied to a variable reaching outside a shared term when this term needs to be made available via sharing from places with a different abstraction context. The correction is simply based on the difference of the depths of the two contexts.

One can see that having the relatively free vars of the more deeply nested *shexp* reaching out less far than the difference of the ϕ's is not within the realm of possibility. The difference of the ϕ's can be positive or negative but its absolute value cannot be larger than the "reach" of the de Bruijn indices of the more deeply nested term.

We now proceed by showing an example of how sharing effectively contracts two or more redices in parallel. This example, in fact, is the one with which Wadsworth illustrated that *RTNF/RTLF* sharing is defied.

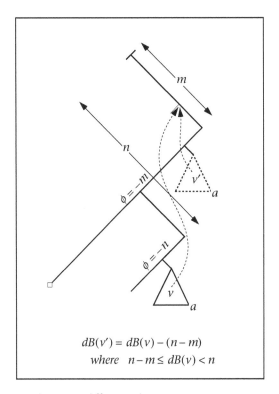

$$dB(v') = dB(v) - (n - m)$$
$$where \quad n - m \leq dB(v) < n$$

Figure 6.1 Shared open subterm in different abstraction contexts.

Table 6.6 shows the equivalent fully-effected reductions of an expression in which the operand subterm is duplicated after the first step. Reduction in isolation of the duplicated subterm (-1 0 ((-1 0) 1)) leads to the contraction of its inner redex thereby saving one reduction. In the detailed trace of Table 6.7 the suspension that is introduced in isolation is underlined (step *D6*). After its reduction is complete (step *R7*), the resulting *shexp* becomes an operator and interacts correctly with the pending argument. Note that the necessary copying of the body of the *shexp* is effected by the action of the *up* procedure in steps *U20-21* forming a portion of the result. On the other hand, the body of *shexp* is included without copying (step *U22*) since the φ in effect is the correct one. *AP-SH* for this expression performs four reductions as shown in Table B.7.

0.	(-2 <u>(-1 0 2 0)</u> (-1 0 ((-1 0) 1)))
1.	(-2 (-1 0 <u>((-1 0)</u> 1)) 1 (-1 0 ((<u>-1 0)</u>....1)))
2.	(-2 <u>(-1 0 1)</u> 1 (-1 0 1))
3.	(-2 1 0 (-1 0 1))

Table 6.6 Equivalent literal substitutions under HOR-SH.

step	term/shexp	stack	env	phi
D0	(-2 (-1 0 2 0) (-1 0 ((-1 0) 1)))	nil	nil	0
D1	(-1 (-1 0 2 0) (-1 0 ((-1 0) 1)))	()	(-1)	-1
D2	((-1 0 2 0) (-1 0 ((-1 0) 1)))	(\ \)	(-2 -1)	-2
D3	(-1 0 2 0)	(([(-1 0 ((-1 0) 1)) ({ -2 -1 })]) \ \)	(-2 -1)	-2
•D4	(0 2 0)	(\ \)	(([(-1 0 ((-1 0) 1)) ({ -2 -1 })]) -2 -1)	-2
D5	(0 2)	(([(-1 0 ((-1 0) 1)) ({ -2 -1 })]) \ \)	(([(-1 0 ((-1 0) 1)) ({ -2 -1 })]) -2 -1)	-2
D6	0	(-1 ([(-1 0 ((-1 0) 1)) ({ -2 -1 })]) \ \)	(([(-1 0 ((-1 0) 1)) ({ -2 -1 })]) -2 -1)	-2
·D7	(-1 0 ((-1 0) 1))	nil	(-2 -1)	2
D8	(0 ((-1 0) 1))	()	(-3 -2 -1)	3
D9	0	(([((-1 0) 1) ({ -3 -2 -1 })]) \)	(-3 -2 -1)	3
U10	0	(([((-1 0) 1) ({ -3 -2 -1 })]) \)		3
D11	((-1 0) 1)	(@ 0 \)	(-3 -2 -1)	3
D12	(-1 0)	(-2 @ 0 \)	(-3 -2 -1)	3
•D13	0	(@ 0 \)	(-2 -3 -2 -1)	3
U14	1	(@ 0 \)		3
U15	(0 1)	()		3
U16	(-1 0 1)	nil		2
R7	(2.0 -1 0 1)	.	.	
D17	(-1 0 1)	(-1 (2.0 -1 0 1) \ \)	(-2 -1)	2
•D18	(0 1)	((2.0 -1 0 1) \ \)	(-1 -2 -1)	2
D19	0	(-2 (2.0 -1 0 1) \ \)	(-1 -2 -1)	2
U20	1	(-2 (2.0 -1 0 1) \ \)		2
U21	(1 0) *copy	((2.0 -1 0 1) \ \)		2
U22	(1 0 (-1 0 1))	(\ \)		2
U23	(-1 1 0 (-1 0 1))	()		-1
U24	(-2 1 0 (-1 0 1))	nil		0

Table 6.7 Reduction of wads34 = (-2 (-1 0 2 0) (-1 0 ((-1 0) 1))) under HOR-SH-VA.

Abstraction-context tagging was invented by Berkling. It is employed in all the incarnations of Head Order Reduction that incorporate sharing while using de Bruijn indices and *UVCs*. We remind the reader that an implementation does not need to actually construct the dummy environment; a special numeric tag embedded in the "real" environment structure is sufficient.

6.4 VARIATIONS OF HOR

This section revisits variations of HOR which were discussed previously under the new context of sharing; it also introduces some new ones. Section 7.4 lists the results of the comparative evaluation of concrete versions which include certain combinations of these variations.

HOR-SH with Early VA

HOR-SH without early *VA* exhibits large inefficiencies for some expressions with little or not potential gain from sharing. The inefficiencies are even more pronounced if one of the other optimizations of *HOR-SH*—the one given the code name *"NEW"* in the following—is also absent.

The overhead of *HOR-SH* without early *VA* is exhibited by using a simple example like *tt1* which does not present any potential for sharing. Under pure *HOR* two reductions are effected and the traversal is a smooth down-up sequence:

D0 D1 D2 D3 D4 D5 D6 •D7 •D8 D9 D10 D11 U12 U13 U14 U15 U16 U17 U18 U19

Under *HOR-SH,* while no savings in reductions are materialized, the down-up motion is interrupted three times for computing two trivial single variable *shexps* and for updating the de Bruijn index of a variable argument:

D0 D1 D2 D3 D4 D5 D6 •D7 •D8 D9 D10 D11 D12 U13 R12 D14 U15 D16 D17 U18 R17 U19 U20 D21 U22 U23 U24 U25 U26 U27 U28

The reader is referred to the complete trace for *tt1* via *HOR* and *HOR-SH* without *VA* which appears in Tables B.2 and B.4 respectively. Below we illustrate the interesting portions.

The next two trace fragments show how the problem starts. In step *D4* and *D5* the two suspensions that are pushed onto the stack have single variable terms. The suspensions become delayed bindings in steps *D7* and *D8*. Under *HOR-SH* these suspensions have to be "reduced" in isolation when referenced from head position.

tt1 HOR-SH:

D4:	((-3 2 1 0) 3 2)	(\ \ \ \)	(-4 -3 -2 -1)	-4
D5:	((-3 2 1 0) 3)	((([2 ({ -4 -3 -2 -1 })]) \ \ \ \)	(-4 -3 -2 -1)	-4
D6:	(-3 2 1 0)	((([3 ({ -4 -3 -2 -1 })])]) ([2 ({ -4 -3 -2 -1 })])]) \ \ \ \)	(-4 -3 -2 -1)	-4
•D7:	(-2 2 1 0)	((([2 ({ -4 -3 -2 -1 })])]) \ \ \ \)	((([3 ({ -4 -3 -2 -1 })]) -4 -3 -2 -1)	-4
•D8:	(-1 2 1 0)	(\ \ \ \)	((([2 ({ -4 -3 -2 -1 })]) ([3 ({ -4 -3 -2 -1 })]) -4 -3 -2 -1)	4

Another inefficiency is that under *HOR-SH* these trivial suspensions turn into degenerate *shexps* like (5.0 . 3) below:

tt1 HOR-SH:

U20:	(4 3)	((([0 ({ -5 (5.0 . 3) (5.0 . 4) -4 -3 -2 -1 })])]) \ \ \ \ \)		-5
D21:	0	(@ (4 3) \ \ \ \ \)	(-5 (5.0 . 3) (5.0 . 4) -4 -3 -2 -1)	-5
U22:	0	(@ (4 3) \ \ \ \ \)		-5

U23: (4 3 0)   (\ \ \ \ \)   -
5

When early *VA* lookup is added to *HOR-SH* one vital implication is that the type of
elements of the stack has to be extended to include *shexp*s as shown in 6.10. The presence of
*shexp*s on the stack also implies that shared expressions can participate in reductions. This
possibility introduces a new set of concerns that are discussed below.

$$Stack \quad s := \varnothing \mid c{:}s \mid \phi{:}s \mid r{:}s \mid \lambda{:}s \mid @{:}s \qquad (6.10)$$

With early *VA*, *UVC*s replace the trivial suspensions and the ensuing extra traversals of
the previous example are avoided:

tt1 HOR-SH-VA:

D4:	((-3 2 1 0) 3 2)	(\ \ \ \)	(-4 -3 -2 -1)	-
				4
D5:	((-3 2 1 0) 3)	(-2 \ \ \ \)	(-4 -3 -2 -1)	-
				4
D6:	(-3 2 1 0)	(-1 -2 \ \ \ \)	(-4 -3 -2 -1)	-
				4
•*D7:*	(-2 2 1 0)	(-2 \ \ \ \)	(-1 -4 -3 -2 -1)	-
				4
•*D8:*	(-1 2 1 0)	(\ \ \ \)	(-2 -1 -4 -3 -2 -1)	-
				4

Summarizing, in cases without potential for sharing *HOR-SH*'s conservative traversal
often has the effect of interrupting the traversal to compute degenerate *shexp*s. Overly lazy
constructions can lead to unnecessary allocation of intermediate structures. Addition of early
VA lookup improves the situation considerably but in some cases does lead to a loss of
sharing. A concrete *HOR-SH* reducer with early *VA* appears in Figure A.11.

Shared Expressions as Operands

The complication that early *VA* introduces by admitting *shexp*s on the stack is that
now a reduced and potentially shared subterm may be an operand of a lazy *hnf*. In the
upward traversal when such a *shexp* is encountered it can be incorporated as is if its tag—
which shows the abstraction context under which it was reduced— is the same as the
current one. In case it is not, its relatively free variables will need correction and therefore it
has to undergo the same processing that a *shexp* faces when referred to from a head variable.
The term of the *shexp* is entered with the proper dummy environment after a marker is
pushed onto the stack to signify its null applicative context.

This is one area of *HOR-SH* reduction where improvements can be made. Since the
extra traversal of an operand *shexp* is not strictly necessary from the point of maintaining

coherency—in contrast to the obligatory one for an operator *shexp*— one can imagine that a delayed form of correction may be feasible. Additionally, one of the optimizations discussed next deals with the idea of actually updating the stack entry after the corrective traversal is complete in the hope that any subsequent encounters are going to happen under the most recent context.

Tabulating the Variations of HOR

Table 6.8 groups the most important variations of Head Order Reduction already discussed and introduces four new ones that are of a quite subtle nature. Their justification in terms of soundness and their effect in terms of runtime behavior range widely. Figure A.13 shows concrete code fragments which realize these variations.

Variation	Description
SH	Head expression reduced in isolation, environment updated and the result entered with the proper dummy environment.
VA	Variable argument looked up immediately and added to the reversed spine instead of forming a trivial closure.
NEW	A newly created *shexp* is not entered if the stack is effectively empty.
READY	An existing *shexp* referenced from a head variable is not entered if it has the correct ϕ and if the stack is effectively empty.
SHEXP.UP	A ϕ-corrected *shexp* on the stack which becomes part of the result also updates the stack entry.
SH.SUSP.UP	Reduction in isolation of a suspension and updating of the stack entry as it becomes part of the result.

Table 6.8 Description of possible variations of HOR .

Early *VA* look-up is essential under *HOR-SH* especially if the *NEW* optimization is not present. Without it, reduction in isolation may be triggered solely for trivial suspensions whose term portion consists of a single variable. As we have just seen expressions which do not inherently contain appreciable potential for sharing are hardest hit. The trace for the particular example we utilized is 50% longer than that of plain *HOR* (Table B.4). When early *VA* lookup is allowed, the effort of reduction with sharing for this expression is identical to that of plain *HOR* (Table B.2). Avoiding the formation of the trivial suspensions appears to save a substantial amount of work for most problem expressions. Early *VA* also is instrumental in keeping the length of the process queue under *BF-HOR* short (cf. section 8.4).

But like many other propositions about efficient Lambda Calculus reduction, the last one too should be taken with a grain of salt. Under sharing early *VA* can actually increase the number of reductions because it leads to a decrease of the "conservativeness" of the algorithm (cf. Table 7.6, section 7.3). We believe though that early *VA* is generally a highly desirable optimization.

The variation termed *NEW* is applicable at the stage where the reduction of a head expression in isolation has just returned a *shexp*; the second traversal of the *shexp* can be omitted in a null applicative context. This information is contained entirely in the state of the stack. The traversal can be omitted when the reversed spine is effectively empty; i.e., when it is entirely empty, consists solely of abstraction markers or when it is locally empty. The reversed spine is locally empty if the topmost item is a marker delimiting the suspended evaluation of an argument.

When a potentially shared subterm is retrieved from the environment in subsequent lookups, it has to be traversed—even in the case of no pending applications (i.e., an empty reversed spine)—for its relatively free variables to be corrected. This traversal effectively performs an on-demand "de-sharing" by selective copying. This last traversal needs to be performed only in case the tagging ϕ of the shared term is different from the ϕ of the current *hnf*. This optimization is termed *READY*. We note that in the case of the first traversal after reduction in isolation the ϕ in effect is constrained to be the same as the tagging one—because the correct dummy environment *is* the one defined by the current ϕ.

The last two variations are trickier and, typically, of more limited effect. In the case of *SHEXP.UP* a *shexp* that is to undergo ϕ-correction because it was encountered during upward traversal out-of-context is corrected in isolation first. Then the new *shexp* is allowed to overwrite the original stack cell in the hope that any further occurrences of the same *shexp* are going to be under the new context.

SH.SUSP.UP implements a form of sharing for suspensions in operand position. During reverse traversal a suspension is reduced with a null stack so that its traversal returns. Then the *shexp* is entered with the proper dummy environment and the rest of the reversed spine in order to continue the overall reduction. In case that the applicative context is null, the second traversal can be avoided and the overall reduction can continue its reverse motion.

As an aside we mention here what is our current thinking with respect to data-driven optimizations. Attempts to make special cases consisting of *I* or (-1 0) in operator position more efficient by avoiding reduction completely does not seem to provide appreciable gains. In a sequential implementation the addition of any conditional test is difficult to justify if the condition does not occur with a significant probability.

In a hardware implementation a unique bit pattern can be assigned to *I* and to some other simple combinator patterns. These patterns can be detected concurrently with other processing when reduction approaches head position by performing a shallow look-ahead. There are many such opportunities for optimizations once the state of the abstract machine is accessible in a "concurrent" fashion.

6.5 INCOMPLETENESS OF HOR-SH

We examine now more closely the serious issue of the incompleteness that is introduced via the strong reduction in isolation of a head expression.

Certain combinations of expressions, reduction strategies and desired normal forms (among, weak, head and strong) can lead to non-termination. In Table 6.9 we collect some of the simplest lambda expression that exemplify these problematic combinations. It is an informal attempt and a point of departure in the effort to characterize the degree of incompleteness of *HOR-SH*.

Starting with the expressions of cases (*A*) and (*B*) we see that they do not possess any normal forms. Hence all reducers are vacuously complete with respect to such expressions.

Cases (*C*), (*E*) and (*G*) consist of expressions which have a *whnf*, *hnf*, and *nf*'s respectively and which pose no problems to any of the reduction strategies. Specifically, the expressions of case (*C*) and (*E*) are already in *whnf* and *hnf* respectively, and the expression of case (*G*) has the diverging argument encapsulated in a *whnf* and so presents no problems under an eager strategy. Under a strong normal order strategy the diverging argument is discarded immediately.

We now proceed with the expressions that do exhibit incompleteness. Cases (*D*), (*F*),

Case	term	whnf	hnf	nf	Incomplete Strategies
A.	Δ	×	×	×	—
B.	((-2 0 1) (-2 0) Δ)	×	×	×	—
C.	(-1 Δ)	√	×	×	—
D.	((-2 1) Δ)	√	×	×	AOR
E.	(-1 0 Δ)	√	√	×	—
F.	((-2 0 1) Δ)	√	√	×	AOR
G.	((-2 0) (-1 Δ))	√	√	√	—
H.	((-2 0) Δ)	√	√	√	AOR
I.	((-1 0) (Y (-2 0)))	√	√	√	AOR
J.	((-1 0) Y (-2 0))	√	√	√	AOR, HOR-SH
K.	((-2 0 1) (-2 0) Y)	√	√	√	AOR, HOR-SH
NOTES:	AOR = Applicative Order Reduction, HOR-SH = Head Order Reduction w. Sharing, Δ = ((-1 0 0) (-1 0 0)), Y = (-1 (-1 0 0) (-1 1 (0 0))).				

Table 6.9 Incompleteness of AOR and HOR-SH strategies.

(*H*) and (*I*) fail to terminate under eager strategies due to the existence of the diverging argument even though reduction to the respective weak, head and strong normal forms does not require its evaluation. Strong reduction in the case of (*D*) and (*F*) fails, regardless of the algorithm employed, since these expressions do not possess a *nf*.

Finally, cases (*J*) and (*K*) exhibit the incompleteness of *HOR-SH*. Even though these expressions have normal forms—they both reduce to the identity term, (-1 0)—the reduction in isolation of subterms referred to by a head variable leads to non-termination. Case (*J*) is straightforward and one can confirm by inspection that the reduction of ((-1 0) *Y*) will fail; case (*K*) is more subtle: the reduction of the *Y* in isolation is attempted only after the delayed reduction of ((-2 0 1) (-2 0)) → (-1 0 (-2 0)) is completed. The last term has its head variable referring to the *Y* and thus the mechanism is triggered. Case (*I*) demonstrates that minor structural differences do indeed manage to avoid the mishap.

One should also note the critical difference between cases (*B*) and (*K*). It is the choice of the diverging subterm. The paradoxical combinator *Y* is a solvable term and, because it replicates its argument, is "self-effacing". This is not true for *Δ* which is unsolvable and entirely "self-contained".

Needless to say, eager strategies fail on both cases which are problematic for *HOR-SH*. But this produces little comfort. Our pragmatically-motivated quest for avoiding duplicate reductions, even though still far from perfect, has bitten us. John Field puts it very graphically:

> … There is evidently a subtle interplay among the issues of efficiency, normalizability, and redex sharing. The quandary is then to find a way to edge closer to the brink of optimality without plunging into the abyss of non-normalizability… [Field 90] p2.

Recapitulating, *HOR-SH* sometimes fails to produce a *nf*. It does so not because it attempts to reduce every subterm—like applicative order does, whether it is needed or not—but because it may attempt to reduce a non-normalizable but solvable subterm outside a context which would otherwise render the subterm harmless. The practical repercussions of this situation are beyond our intellectual reach at this time.

Do other, less drastic forms of sharing fit well within *HOR*? We believe the answer is a qualified "yes". Reducing in isolation a head-referred expression only to *whnf* will certainly result in a complete algorithm but also to one with a substantially decreased degree of sharing. Reducing in isolation such an expression to *hnf* could be a better tactic. We would still be avoiding incompleteness while increasing the degree of sharing. *HOR-SH* opts for the most eager option—i.e., reducing to *nf*—a tactic which maximizes sharing but does not manage to avoid incompleteness. More work on integrating all three options in *HOR* and gaining empirical feeling is needed before a more educated judgment can be made.

Chapter 7

Comparative Analysis and Evaluation

7.1 General

It is impossible to assemble a set of test cases for which one can claim that they are representative of all the expressions that may be presented to a lambda reducer. Only if such a set could be devised one would be allowed to attach a figure of merit to a specific reduction algorithm based on its performance on a common test set. The Lambda Calculus can emulate computations of arbitrary complexity and so one could conclude just about any desired conclusion based on results on specially rigged expressions[31]. On the other hand, a test set is of vital importance—at least in our pragmatically-oriented work—for exercising the reduction algorithms we have developed and comparing classes of algorithms to each other.

It is expected to have these difficulties when one is confronted with such a general problem. G. Revesz puts it like this:

> ... We are dealing with the efficiency of the process of evaluating arbitrary partial recursive functions. Standard complexity theory is clearly not applicable to such a broad class of computable functions. The time complexity of such a universal

---

[31] In particular, we conjecture that once a strategy has been fixed one can devise example *wfe*'s that will attribute to that particular strategy a cost that is arbitrarily larger than the theoretically minimum one arrived to by counting "parallel" reductions; see for instance some of the constructions by Frandsen and Sturtivant [Frandsen 91].

procedure cannot be bounded by any computable function of the size of the input. ... [Revesz 88].

Instead of attempting to be fair when it is known that such an attempt is futile and in view of limited resources we tried to prevent extreme bias by employing a simple, but far from safe, tactic: the set of test cases that are included in this dissertation as the *all** set is one that was brought into bear over time by the continuous and long-lived process of looking for challenging cases[32].

In addition to the main set of test cases which attempts to be fair by lack of design, several specialized sets designed to test specific aspects of the reduction algorithms were also devised. These typically consist of expressions exhibiting certain patterns or which apply an operator to arguments of increasing size. In this chapter we will attempt to share the experiences we gained from experimentation with the various reducers.

The measurements for time and memory usage are provided by an experimental system. In our setup before the reduction of a test expression is initiated we note the system time and the total size of memory already allocated. After reduction is complete we take new measurements. The respective differences are the estimates of the total time and of the total size of the structures allocated for reducing an expression and forming a result. Early on, some hand calculations for memory usage were done. We verified that the reported numbers are accurate.

The reported estimates of time are those reported by the built-in functions of *Macintosh Common Lisp* running on a 68030 processor at 25MHz. Accuracy is limited to about one millisecond. Time measurements are consistent and is some cases were made even more accurately by temporarily disabling interrupts. Throughout, memory measurements are in bytes (alternatively, *octets*) and time measurements are in seconds.

As mentioned in section 2.2 the "HOR-style" graph representation is actually quite wasteful (cf. Table 2.1). This point will be reinforced in the discussion of *μRED* in the next chapter. We did not attempt to take measurements using other representations except in the case of the applied simulator/reducer, *BETAR3*. It is safe to state that memory consumption has been estimated conservatively.

To refresh the reader's memory on the names of the various reducers it is helpful to refer to Table 1.6 on page 18. The *VA* suffix means that a version of the reducer with the early variable argument lookup optimization was employed. The *SH* suffix means that a reducer incorporates sharing.

---

[32] The set *all** is listed in Appendix D.

Summarizing, it is easy to question the wisdom of judging the merits of reduction algorithms by looking at families of hard or even pathological test cases[33]. But one also has to admit that sometimes this is all one has to resort to when dealing with rich theories such as the Lambda Calculus where a reduction algorithm practically has to negotiate, in a fundamental manner, all expressible computations. The findings of this chapter should be read under this spirit.

7.2 COMPARISON OF WEAK REDUCERS

The weakly normalizing reducers present difficulties in terms of devising meaningful tests in the context of the pure calculus because many of the candidate expressions are either in *whnf* or reach *whnf* quickly after a few reductions. One has to find tests that allow the extraction of substantive operational information and this means that they must include some form of self-application.

One test which is well suited to all classes of reducers is the Church Numeral predecessor function *pred*. The definition of *pred* has already appeared multiple times (cf. section 1.5, Table 1.4). An annotated reduction sequence for *pred2* can be found in Table B.1.

Reducer	Reds.	Calls	Memory	Time
W-HOR-VA	447	1 213	14 112	0.061
RTLF-VA	447	1 213	24 840	0.076
W-HOR-SH-VA	98	307	3 920	0.018
RTLF-SH-VA	98	270	5 384	0.021
SECD-V	22 952	114 762	2 677 528	8.26
EV-AP	22 952	91 809	839 384	3.44

Table 7.1 Results of whnf reduction for pred7 .

Table 7.1 shows the runtime statistics of six *whnf* reducers on the *pred7* test. One observes that for this test expression applicative order proves to be a grossly inferior strategy—even from the point of economy. This is the reason why *π-RED*, the eager system from the University of Kiel, does not do so well in the *pred** tests (cf. section 7.6).

When sharing is included it is notable that the best weak reducers take 98 reductions to produce a *whnf* while the fully normalizing *HOR-SH* produces a *nf* with 80 reductions. It

---

[33] Wadsworth showed that for any pair of integers m, n with $2 \leq m \leq n$ one can find a *wfe* for which graph and string normal-order reduction will have lengths m and n, respectively [Wadsworth 71], p. 176. He also commented that even though all combinations are theoretically possible there is merit to *statistical* comparisons. And he did not fail to mention their inherent difficulty.

does so though at a higher real cost (Table 7.3). *W-HOR-SH* shares the behavior of the *RTLF* strategy. *AP-SH-VA* does very well on this test, costing only slightly more than *RTLF-SH-VA* while producing a full *nf*.

W-HOR owhnf after 447 reductions
([(-2 2 1 (1 0)) ({ ([([(1 (1 (1 (1 (1 (1 0)))))) ({ ([(-4 1) ({ ([(-2 1 (1 (1 (1 (1 (1 (1 0)))))) ({ })]) })]) }) ([(-1 0 1 (-2 0) (-2 0) (-2 2 1 (1 0)) ({ ([(-2 1 (1 (1 (1 (1 (1 (1 0)))))) ({ })]) })]) })]) ([(-2 1 (1 (1 (1 (1 (1 (1 0)))))) ({ })]) })])

W-HOR-SH owhnf after 98 reductions
([(-2 2 1 (1 0)) ({ ([(-2 2 1 (1 0)) ({ ([(-2 2 1 (1 0)) ({ ([(-2 2 1 (1 0)) ({ ([(-2 2 1 (1 0)) ({ ([(-2 2 1 (1 0)) ({ ([(-2 0)({ ([(-4 1) ({ ([(-2 1 (1 (1 (1 (1 (1 (1 0)))))) ({ })]) })]) ([(-2 1 (1 (1 (1 (1 (1 (1 0)))))) ({ })]) })]) })]) ([(-2 1 (1 (1 (1 (1 (1 (1 0)))))) ({ })]) })]) })]) ([(-2 1 (1 (1 (1 (1 (1 (1 0)))))) ({ })]) })]) ([(-2 1 (1 (1 (1 (1 (1 (1 0)))))) ({ })]) })]) ([(-2 1 (1 (1 (1 (1 (1 (1 0)))))) ({ })]) })]) ([(-2 1 (1 (1 (1 (1 (1 (1 0)))))) ({ })]) })])

Table 7.2 Operational *whnf* results of the pred2 test.

W-HOR and *W-HOR-SH* (and also *RTLF* and *RTLF-SH*) produce operational *whnfs* with differing environments (Table 7.2). As expected, the environment portion of a result *whnf* produced by the sharing versions is less "ready" or more lazy. The term portions of both *whnfs* are identical. The two *owhnf* results of course are equivalent if the indicated substitutions are fully effected.

Reducer	Reds	Calls	Memory	Time
HOR-SH-VA	80	713	10 584	0.037
AP-SH-VA	110	320	6 352	0.024

Table 7.3 Results of nf reduction with sharing for pred7.

The next two tables 7.4 and 7.5 show that the weak reducers, especially the two eager ones, leave much room for improvement on this test. In the first one we see that *HOR* and *AP* strong normal order reducers can produce full normal forms for this expression at about double the cost of their weak counterparts. In the second table we see that, under sharing, a much more difficult problem—computing the predecessor of the 14th Church Numeral— requires resources still well under those expended by *SECD-V* and *EV-AP* on *pred7*.

Reducer	Reds	Calls	Memory	Time
HOR-VA	878	2 407	28 008	0.133
AP-VA	878	2 399	49 096	0.148

Table 7.4 Results of nf reduction without sharing for pred7.

Reducer	Reds	Calls	Memory	Time
HOR-SH-VA	255	2 526	39 072	0.133
AP-SH-VA	411	1 167	22 424	0.080

Table 7.5 Results of nf reduction with sharing for pred14.

Head-normal reduction for this test produces results nearly indistinguishable from those of strong reduction. *HN-HOR* requires the same number of reductions and incurs nearly all the cost of full normalization.

We are not going to elaborate any further on the performance of the weak reducers. We presented a comparison of the weak reducers on a single test expression only as a prelude to the main focus of this chapter, the comparative exercise of the strong reducers.

7.3 REDUCTION COUNTS AND COSTS

Adopting a Cost Metric

A first concern is to define a gross but meaningful measure of the actual costs of reduction. The vital observation that leads to our decision is the strong correlation of required time and memory resources.

Figure 7.1 plots these measurements for three of the abstract reducers with and without sharing on the *all** set of test expressions. Figure 7.2 on the next page plots the same information for the *BETAR3* simulator. All plots use a logarithmic scale for both memory and time.

Since there is a strong correlation between memory utilization and runtime it makes sense to assign a single scalar metric for the cost of normalizing an expression. We define

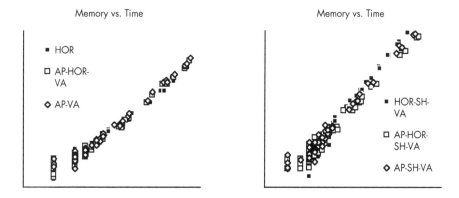

Figure 7.1 Memory vs. time plots of three abstract reducers (all* tests).

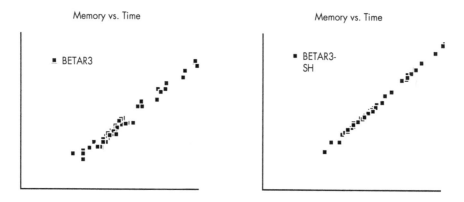

Figure 7.2 Memory vs. time plots of the BETAR3 simulator (all* tests).

such a metric C to be the geometric mean of total memory and time resources.

$$C = \sqrt{m \cdot t}$$
$$where \quad m = \quad \text{allocated memory} \quad (octets)$$
$$t = \quad \text{elapsed time} \quad (seconds)$$

The absolute value of this metric is entirely dependent on the specific implementation and runtime environment; nonetheless it is quite valuable for relative efficiency measurements. The specific values reported in this chapter are those produced by our experimental setup.

Reduction Counts

We begin our comparisons by focusing on reduction counts. Figure 7.3 plots on a logarithmic scale the results of the 55 tests of the *all** set. Tests are listed in order of increasing reduction counts. One can observe the following:

- The reduction counts of pure normal-order reduction are an upper bound of the reduction counts of each test. The sharing versions, *HOR-SH* and *AP-SH* perform fewer reductions than those of pure normal order[34]. The non-sharing versions of strong *nf*

---

[34] We believe that Wadsworth's proof of Normal Graph Reduction (*NGR*) applies in the case of delayed methods like *AP-SH*. We are not certain whether its premise—i.e., that the pure normal-order reduction counts constitute an upper bound—holds for *HOR-SH*. Even though statistically *HOR-SH* is more conservative than *AP-SH* in terms of reduction counts we suspect that there are expressions for which *HOR-SH* requires more reductions than *HOR*.

reducers always do exactly the same number of reductions, i.e., the one that is defined by literal leftmost-outermost substitutions.

- Aggregate results indicate that *HOR-SH* tends to perform substantially fewer reductions than *AP-SH*. The totals are tabulated in Table 7.6 below. Additionally, *HOR-SH* performs fewer reductions for reaching normal form for every test in the *all** set except one (*pones*; 22 vs. 15).

- Early *VA* lookup optimization can increase reduction counts even while lowering total reduction costs (Table 7.6). Under *AP-SH* the only tests whose reduction counts are affected are the *ashar** ones but nearly all tests benefit in terms of reduction cost. Similarly, under *HOR-SH* only the *tmh9* and *tw10-2* tests show an increase but nearly all tests benefit in terms of cost. Early *VA* appears to lower costs by about 15% over all the tests of the *all** suite for both *HOR-SH* and *AP-SH* but the effects on individual tests vary widely.

- The test with the most dramatic ratio between the sharing reducers is *fs4n8*[35]. *AP-SH* fails to show any benefit over normal order and exhibits an exponential count of reductions as a function of the problem size. *HOR-SH* on the other hand reduces enough in "isolation" to bring the count down to linear. Applicative order also behaves linearly. This family was described by Frandsen and Sturtivant [Frandsen 91] as an example of the exponential inefficiency of combinator reduction in relation to direct lambda calculus reduction.

We did not attempt to measure reductions under a strong applicative order strategy. We do know that the *pred7* and *pred9* tests are entirely impractical[36] and that *nbot* would fail to terminate.

Reductions	Normal-Order	AP-SH	HOR-SH
without VA	13 147	5 546	753
with VA	13 147	5 719	769

Table 7.6 Total reduction counts for all* with and without early VA.

---

[35] This expression is the eighth member of the set with elements $A_0 = \lambda x.I$, where $I = \lambda x.x$ and $A_n = \lambda h.((\lambda w.wh(ww))A_{n-1})$.

[36] *pred7* by itself would require more reductions than those expended for the whole set *all**.

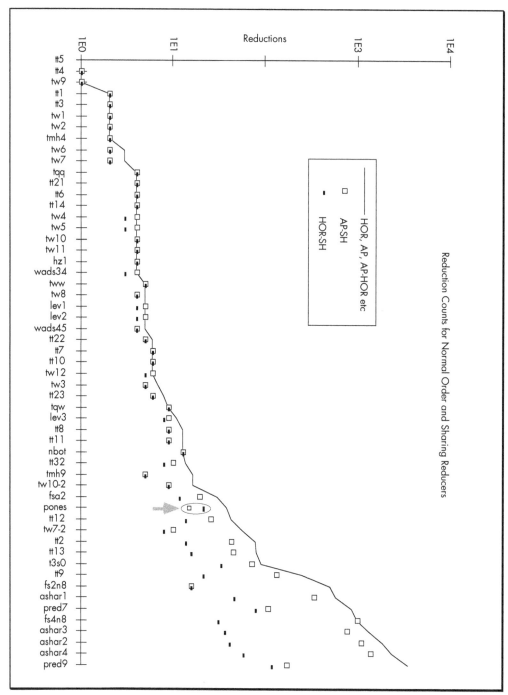

Figure 7.3 Reduction counts for normal-order and the two sharing reducers for the all* test suite.

Reduction Costs

Next we focus on reduction costs. Figure 7.4 plots on a logarithmic scale the reduction costs of the 55 tests of the *all** set. Individual tests are ordered by increasing average reduction costs under three of the pure normal order reducers. We observe the following:

- The reduction costs of the *HOR, AP* and *AP-HOR* pure normal-order reducers have a good correlation. The logarithmic scale helps of course in de-emphasizing fractional differences. The general conclusion is that minimal normal-order reducers are nearly identical in this respect. A related plot was previewed in section 4.6.

- Sharing reducers tend to exhibit dramatic improvements in reductions costs for the tests that have high enough costs and which have the potential to benefit from sharing.

- It is not the case that the reduction costs of pure normal order reducers constitute an upper bound on the costs of the sharing reducers. The sharing version of *HOR* often exceeds in cost the pure ones. Simple *wfes* with few reductions do not, as a rule, benefit from sharing because the effort expended in arranging for sharing does not compare well with respect to the final gain.

- The last attribute of *HOR-SH* is more pronounced on the left side of the plot. *HOR-SH* appears to often expend more effort than normal reduction without sharing or *RTNLF* reduction with sharing for producing a normal form for the tests whose demands are modest. It can easily be explained by the more conservative attitude of the reduction in isolation; it provides substantial benefits in situations where subsequent work is avoided but not in the cases that do not include enough sharing potential. *AP-SH* manages to avoid this overhead and appears to have respectable average performance. But it also fails to exhibit benefits from sharing on some tests (*fs4n8*) and does not do as well on some others (*ashar**).

These empirical tests show that the runtime behavior is mostly governed by the main strategy. Representation and optimizations statistically do not have as much of an effect in reducers which have a careful and minimal specification. But as we shall see, optimizations that do not seem to have an appreciable overall effect sometimes influence the runtime behavior greatly in specific tests.

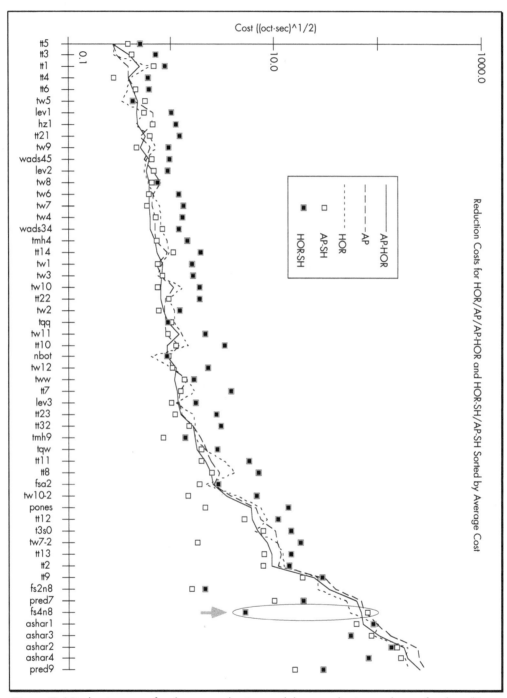

Figure 7.4 Reduction costs for three non-sharing and the two sharing reducers for the all*
test suite.

Figure 7.5 shows on a logarithmic scale the relative costs of *HOR-SH* and *AP-SH* from another perspective. It plots the ratio of reductions and the ratio of costs together with the spread between the ratios of memory utilization and runtime. The tests are sorted in order of increasing cost ratios. In this graph one can observe the following:

- The ratio of the costs and the individual ratios of memory and runtime (height of the lines superimposed on each point) correlate well with the exception of a few tests.

- The plot appears to have relatively distinct parts. For most expressions in the *all** set the cost ratio remains in the range of 1:2 or 1:3 in favor of *AP-SH*. On the right extreme of the curve are few tests that show ratios that are significantly larger than the those of the middle range.

- The overall trend is for *AP-SH* to be more efficient but to also perform more reductions. There are three tests which require more reductions under *HOR-SH* than under *AP-SH*.

- The tests that exhibit significant ratios in favor for *HOR-SH* are those which also show large reduction ratios in favor of *HOR-SH*. In these tests *HOR-SH* achieves an appreciably higher degree of sharing.

- *fs4n8* is the extreme case where no benefit is shown by the sharing strategy of *RTNF/RTLF* and all the benefit of applicative order is achieved by the sharing strategy of *HOR*. The result is a reduction from an exponential to a linear complexity.

Again, the reader is cautioned that these are empirical observations on a relatively small number of tests and they may or may not be indicative of more "general attributes"— if such notion exists for Lambda Calculus reduction.

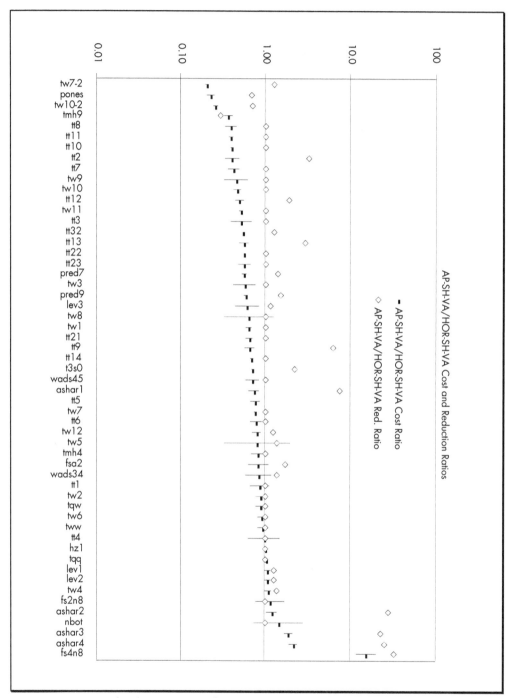

Figure 7.5 Cost and reduction ratios of AP-SH vs. HOR-SH for the all* test suite.

Reducing/Checking of Normal Forms

Figure 7.6, is a logarithmic plot of the *ratio* of the costs for the respective members of *all*/rall** sets. It is included in support of the premise which has been stressed before: in many cases the costs of normalization are comparable with the costs of traversal performed with an "intention for reduction". The effect of "intention" is important in light of the discussion that follows. Meanwhile we observe the following:

- Only a small percentage of tests exhibit ratios larger than one order of magnitude. Expressions which have appeared in the Lambda Calculus literature as examples of reduction with varying degrees of non-optimality show ratios between one and two. This means that elaborate sharing mechanisms often do not produce cost benefits that exceed the costs expended to allegedly achieve them.

- A cost ratio is high either because a reducer is inefficient in reducing a expression or because it is uncommonly efficient in dealing with its normal form. Judging from the specific tests under *AP-SH-VA* that show relatively large ratios compared to those of *HOR-SH-VA* we believe that the former possibility is more often the case. Relative efficiencies with respect to expressions already in normal form usually are of a small and relatively constant ratio whereas in this figure we see some quite pronounced differences (*fs4n8*, *nbot*, *ashar3**, etc.)

- A cost ratio is also affected by the *size* of the test expression and its normal form. Expressions which pack little work in relation to their initial or final size will tend to exhibit ratios close to one because the costs of traversal dominate the costs of reduction.

- The range of ratios is definitely a property of the set of the expressions comprising the test suite rather than a general trend. But we think this does not invalidate the experiment. The premise above should not be construed as an assertion that a large amount of work cannot be demanded by an expression with a compact normal form. That would be ludicrous. It should instead be construed as an indication that, statistically speaking, the overall cost of reduction is strongly influenced by the cost of traversal. Tests, like those of the *fsa** family (see section 7.6), can be made to contain an arbitrarily high cost ratio but the reader will agree that there is something contrived about these tests when they are viewed from a computing point of view. This realization leads us to state that attempts for quantifying the cost of reduction in terms of the characteristics of the problem expression should definitely include size metrics.

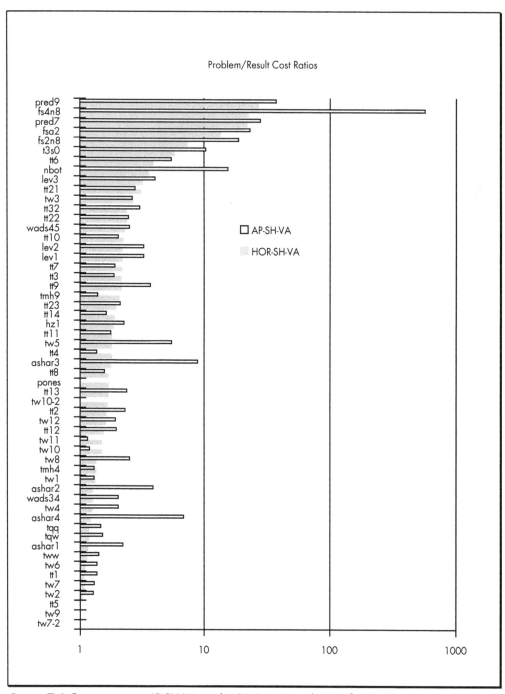

Figure 7.6 Cost ratios via AP-SH-VA and HOR-SH -VA reduction for problem-result pairs.

Problem/Result Cost Ratios

One should realize that the intent of an algorithm to check for normal form or to perform reductions is of consequence. Just checking whether an expression is in normal form does not require a full-fledged reducer. But efficient reduction procedures cannot be excessively burdened in comparison to a minimal algorithm that simply tests for the absence of a redex.

In this spirit, Figure 7.7 plots the ratio of the costs of *BF-HOR, HOR* and *AP* to the cost of a minimal procedure that tests for normal forms. The horizontal axis of this plot are the individual members of the *rall** suite which are expressions in normal form consisting of the fully reduced members of the *all** test set.

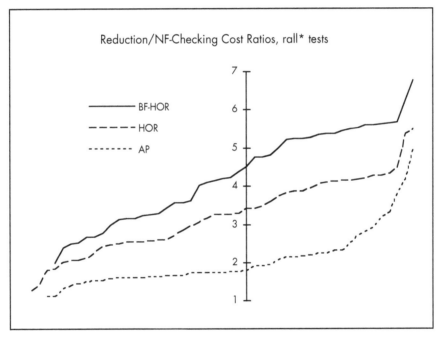

Figure 7.7 Ratios cost of "reduction" vs. nf checking for the rall* test suite under BF-HOR, HOR and AP reduction.

We see that the three reducers exhibit ratios in the range of roughly two to seven. This is considered reasonable given that reduction algorithms "remember" the structure of the problem expression and have to make entries for the free variables in the environment (which increases space) and do more branching (which increases time). As expected from previous discussion, *AP* has the least overhead and *BF-HOR*, with its increased environment residency, the most.

For completeness, the algorithm which checks expressions for normal form is sketched next. The only critical observation needed is that an expression in *nf* is also in *hnf*. Therefore an algorithm which traverses the expression can rely on this fact and immediately return a negative answer if a redex is found along the left spine. We call this procedure *nfcheck*.

Procedure nfcheck (term)

A single depth-first traversal suffices. Top-level abstractions of the term are skipped and subsequent operands are collected on a stack until a head variable is reached. Each of the operands is recursively *nfcheck*ed after exchanging it on the stack with a special marker. If at any point an abstraction is encountered while the top of the stack is anything but a marker the answer is *no*. If we ascend fully and the stack is empty the answer is *yes*. If the term is finite this test always terminates with an answer.

When confronted with expressions already in *nf*, the sharing properties of a reducer are not relevant. Since no reductions are done, all environment lookups result in *UVCs* or unbound Aiello–Prini dummy variables. At no time is there opportunity for reduction "in isolation" of an operator or a subterm. Under both *HOR* and *AP* reduction the sharing versions exhibit no overhead. This is the reason only three families are shown in the last figure. Early *VA* lookup on the other hand appears to make a small difference by simplifying the processing of variable arguments.

Overall, measurements indicate that the *HOR* strategy is slightly costlier than the *RTNF/RTLF* one in determining whether an expression is already in normal form.

Church Numeral Predecessor Tests

The next series of experiments involve again the predecessor operator for Church Numerals. Reducing expressions of this test family to either *whnf*, *hnf*, or *nf* is very demanding. Naive strategies result in a high degree of redundant work. Such behavior is exhibited under both pure normal order strategies and applicative ones (Figure 7.8). The normal order sequence of reductions for *pred2* is shown in Table B.1.

How well do the more efficient reducers do? *HOR-SH* performs the fewest reductions (Table 7.7). The *RTNF/RTLF* strategy seems better in terms of overall costs though. The ratios for all three measures remain close to 50% (Figure 7.9). We believe ratios of this magnitude can be nullified in favor of *HOR-SH* under a low-level implementation. Neither of these sharing reducers is strictly tail-recursive but *HOR-SH* includes the reversed-spine stack explicitly whereas *AP-SH* will require the addition of at least one stack to manage the recursive calls on operator subterms.

Test	Normal-Order	AP-SH	HOR-SH
pred1	8	8	8
pred2	20	15	15
pred3	46	26	24
pred4	100	41	35
pred5	210	60	48
pred6	432	83	63
pred7	878	110	80
pred8	1 772	141	99
pred9	3 562	176	120
pred10	7 144	215	143
pred11	14 310	258	168
pred12	28 644	305	195
pred13	57 314	356	224
pred14	114 656	411	255
pred15	229 342	470	288
pred16	458 716	533	323

Table 7.7 Reduction counts for pred1-16.

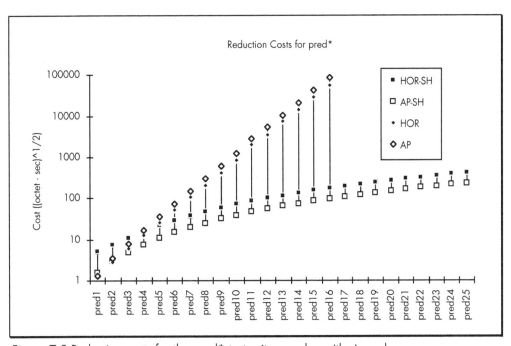

Figure 7.8 Reduction costs for the pred* test suite on a logarithmic scale.

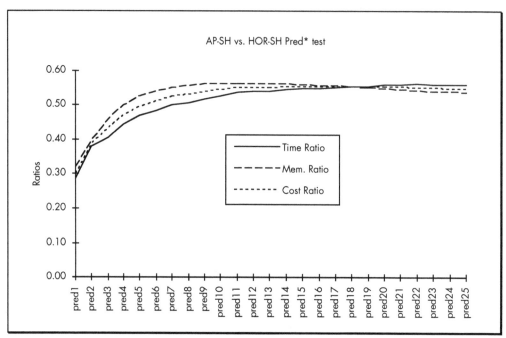

Figure 7.9 Ratios of AP-SH/HOR-SH for the pred* test suite.

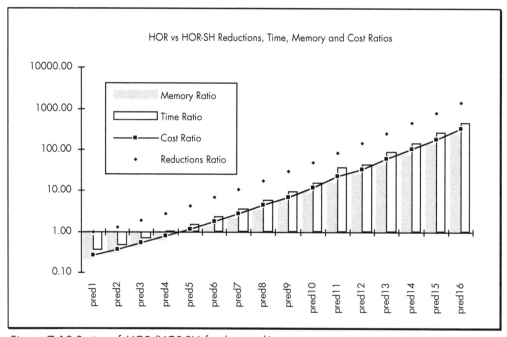

Figure 7.10 Ratios of HOR/HOR-SH for the pred* test suite.

Figure 7.11 shows memory utilization and reductions plotted against time for the four combinations of *HOR* and *AP* with and without sharing. The correlation of memory and time is again excellent. From this and the previous figures one can make the following observations:

- The non-sharing abstract implementations exhibit exponential complexity[37]. Curve fitting for memory, reductions, and time produces formulas of the form $k \cdot e^{0.7n}$. Correlation is excellent.

- The sharing versions succeed in compressing all the above measures to quadratic complexity. The number of reductions as a function of n drops to $n^2 + 4n + 3$. Similar complexities are obtained for memory utilization and time. The *pred** tests have not been analyzed any further to ascertain if this is indeed the theoretical minimum in the sense of Lévy.

- Pure *HOR* is again slightly better than pure *AP*; the improvement is due to better memory utilization.

- *HOR-SH* shows an overhead on both time and memory compared with *AP-SH*. The latter seems to achieve the benefits of sharing without incurring additional costs (observe the collinearity of *HOR-SH*, *AP* and *AP-SH*).

- For this family, *HOR-SH* is approximately 55% more expensive than *AP-SH* (cf. Figure 7.9). Due to its overhead, *HOR-SH* does not improve on *HOR* until after the first few members of this series (cf. Figure 7.10).

- The non-sharing versions perform identical number of reductions; i.e., exactly those required by the normal order strategy. In the sharing versions, *HOR-SH* again tends to do about $1/3$ fewer reductions than *AP-SH*.

- The mere fact that with the sharing reducers one can compute on a personal computer up to *pred25* and beyond is quite remarkable. The reader is invited to review section 7.6 where results of the same test under the *HORSE* and π-*RED*s systems are reported.

- *AP-SH* appears to be the most nimble reduction algorithm on the expressions of our main test set but it fails to show any benefits from sharing in some cases. *HOR-SH* tends to be more "conservative" and in general performs fewer reductions. But this comes at the cost

---

[37] This is also true of *HORSE* and π-*RED* as we shall see in section 7.6.

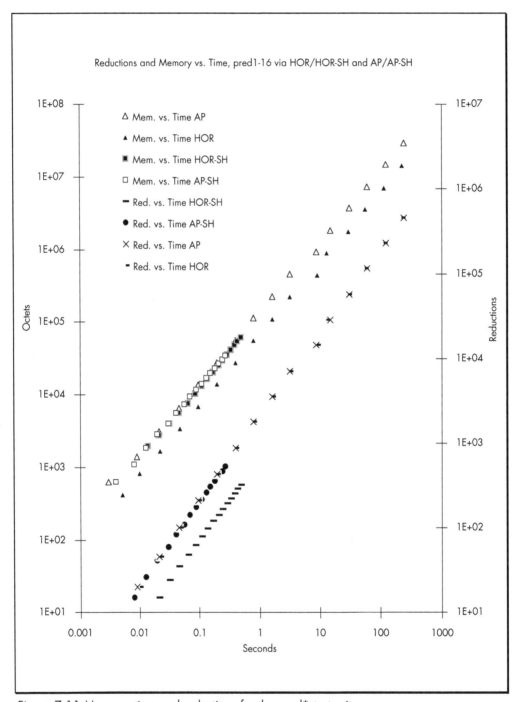

Figure 7.11 Memory, time and reductions for the pred* test suite.

of rescanning which is not justified if subsequent possibility for sharing is not materialized due to the particular structure and dynamics of the expression.[38]

Summarizing, these experiments give one the feeling that the optimal reduction sequence cannot be approximated without a lot of inspection and bookkeeping. Strategies like *RTNF/RTLF* and *HOR-SH* appear to be interesting compromises. Applicative order, in light of the results in the beginning of this chapter and on the *pred** tests does not appear as meritorious—even from the point of economy. Techniques which attempt to lower reduction counts further are expected to have increasingly higher levels of overhead.

7.4 EVALUATION OF VARIATIONS OF HOR

The difficulty of the task of assigning merit to a reduction algorithm can be appreciated even more when one examines the range of costs of reduction of a common test set under variations of the same basic strategy. This section tabulates the results of reducing the *all** suite under fourteen combinations of variations of Head Order Reduction. The cumulative costs of the reduction under each of the combinations is listed in Table 7.9. For reference we repeat the definition of these variations (cf. section 6.3) in Table 7.8.

Under most cases the effect of combining variations of *HOR* is additive in the sense that they influence the total cost in the same direction and by a predictable amount. In other cases this is not true. We have to stress again that the absolute values obtained for the reduction costs are unimportant and are influenced excessively by the presence of tests whose reduction is much more demanding than that of the "average" test in the set. But it is also true these results expose weaknesses of reducers which lack certain optimizations. For example, the presence of the *VA* or *NEW* optimizations is by itself enough to avert a substantial (nearly ten-fold) increase in the costs of the *ashar** tests. We refer to the *SH* and *SH-READY* combinations. They have been omitted from the graph of Figure 7.12 in order to avoid making the range of its cost scale even larger.

The most important aspect of this exercise is of course the reaffirmation of the benefits of sharing which, for this test suite, result in halving the costs of pure normal order. But we observe that *VA*, *NEW* and *READY* should not be just considered optimizations but necessary additions to the sharing versions of *HOR*.

---

[38] There is also the potentially more serious issue of non-termination which was discussed in section 6.5.

Variation	Description
SH	Head expression reduced in isolation, environment updated and the result entered with the proper dummy environment.
VA	Variable argument looked up immediately and added to the reversed spine instead of forming a trivial closure.
NEW	A newly created *shexp* is not entered if the stack is effectively empty.
READY	An existing *shexp* referenced from a head variable is not entered if it has the correct ϕ and if the stack is effectively empty.
SHEXP.UP	A ϕ-corrected *shexp* on the stack which becomes part of the result also updates the stack entry.
SH.SUSP.UP	Reduction in isolation of a suspension and updating of the stack entry as it becomes part of the result.

Table 7.8 Definition of variations of HOR.

No.	Cost‡	Mix of HOR Variations
0.	609.0	*MIN†*
1.	648.8	*SH-VA-RDY-NEW-SXP.UP*
2.	**651.6**	**SH-VA-READY-NEW**
3.	654.0	*SH-VA-NEW*
4.	659.4	*SH-VA*
5.	661.0	*SH-VA-READY*
6.	713.2	*SH-NEW-READY*
7.	732.2	*SH-NEW*
8.	1188.6	*SH-VA-SH.SUSP.UP*
9.	1261.3	*VA*
10.	1355.0	*SH-SH.SUSP.UP*
11.	**1585.3**	**HOR**
12.	1663.1	*VA-SH.SUSP.UP*
13.	3979.6	*SH*
14.	4010.6	*SH-READY*

‡ $(octet \cdot second)^{1/2}$
† Selection of the best measurement for each test among all reducers 1-14.

Table 7.9 Cost statistics of fourteen combinations of HOR on the all* suite.

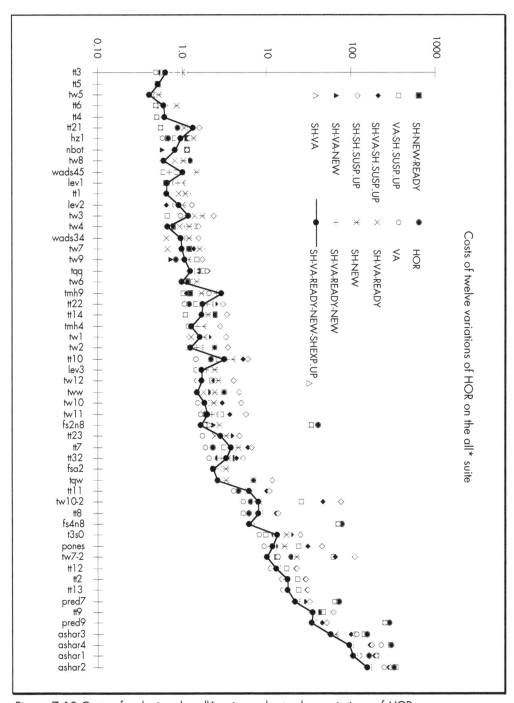

Figure 7.12 Costs of reducing the all* suite under twelve variations of HOR.

7.5 TESTS OF RECURSION & SHARING

For this series of tests we employ the *BETAR3* simulator, a more detailed operational version of *HOR* for an applied Lambda Calculus[39]. The runtime costs of this simulator are not directly comparable with those obtained by the more abstract versions because of the different coding style and the effects of more detailed instrumentation (cf. section 7.7).

An applied λ-calculus simulation of the *factorial* function exemplifies in the most direct manner the necessity of including sharing in any implementation with claims to competence. Inclusion of sharing is by itself enough to lower the complexity of the factorial function to linear (Figure 7.13).

Sharing also permits the inclusion of a "loop-forming" rule for the Y combinator. The last combination manages to cut the runtime in half while reducing the number of reductions slightly (Figure 7.14). After the constant and linear components are subtracted the remaining non-linear component of the costs of the non-sharing versions is verified to be of the second degree. The number of reductions is linear on the size of the input for both sharing versions at $7+6n$ and $9+7n$.

Figure 7.15, a magnification of the initial portion of Figure 7.14, shows again that sharing under *HOR* incurs extra costs. In the case of the *tfac** tests without the cyclic rule for Y, the overhead is not exceeded by the gains of sharing until the ninth test. When the cyclic rule is included the overhead of the extra machinery is minimal.

Figure 7.16 plots on a logarithmic scale the runtime of the *Fibonacci* tests. The effect of sharing is evidenced by the decrease of the polynomial complexity of this family by one order; witness that the slopes of the sharing versions are smaller.

Use of the Y combinator to effect recursion is similar to the *labels* special form of *Lisp* [McCarthy 60].

---

[39] "Applied" in the sense of including primitives for arithmetic, recursion, syntactic term equality, etc. In addition, the *BETAR3* simulator has as its goal to simulate in some detail the procedural aspects of a hardware *HOR* machine.

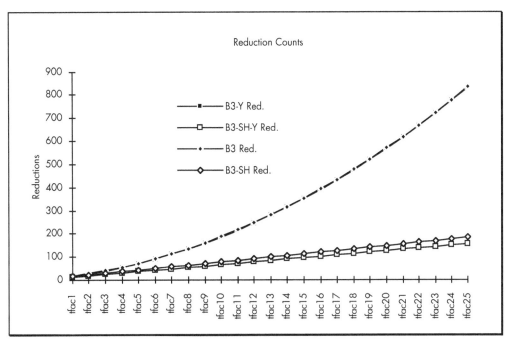

Figure 7.13 BETAR3 reduction counts for the tfac* test suite.

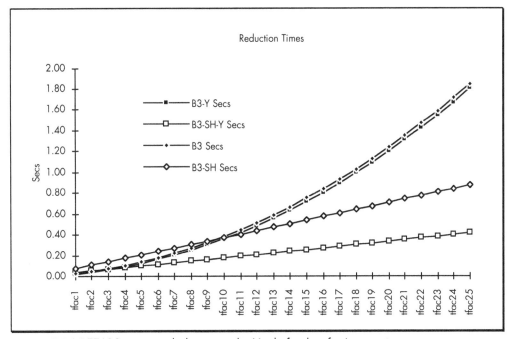

Figure 7.14 BETAR3 times including a cyclic Y rule for the tfac* test suite.

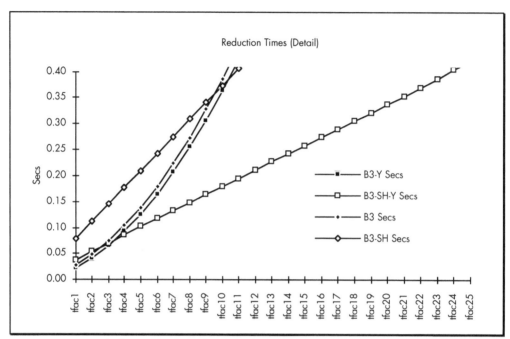

Figure 7.15 Detail of BETAR3 times for the tfac* test suite.

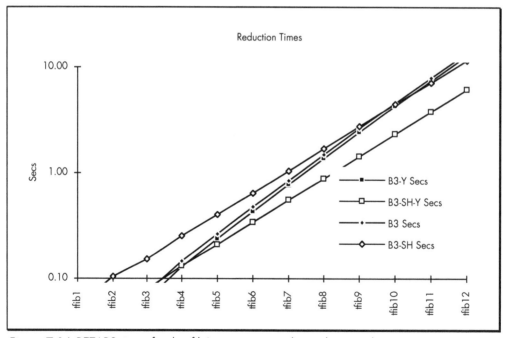

Figure 7.16 BETAR3 times for the tfib* test suite on a logarithmic scale.

– 152 –

7.6 COMPARISONS OF HORSE AND π-RED

In this section we present the results of a few measurements on three expression-based reduction systems that run under Unix on SUN workstations. While these systems are not yet of "industrial strength" they are fairly complete implementations that can be useful in application work. These systems are the π-RED* and π-RED+ from researchers at the University of Kiel [Schmittgen 92] [Kluge 93], and the *HORSE* system by Michael Hilton developed at Syracuse University [Hilton 90a] [Hilton 90b]. The π-RED* and π-RED+ systems use applicative-order; π-RED* reduces only to *whnf*. *HORSE* is normal-order and reduces to *nf*.

Comparisons of these systems are useful because they all have low-level implementations in the *C* language and because they are "applied" systems with primitives operations and data structures and therefore provide some indication how these additions affect raw performance. Such measurements can also put the numbers obtained from the abstract reducers under better perspective.

The reader is reminded that unless otherwise noted the times reported in this section are those for a SUN server 4/490 under SUNOS 4.1.3. In the tests of the abstract reducers the times reported are for *Macintosh Common Lisp* 2.0 on a 25MHz 68030 processor. The latter platform is estimated to have 15-30 times lower raw performance than that of the server. It is not attempted to give normalized numbers for time because the multitude of interfering factors do not permit us to assign a single scalar as an accurate overall performance ratio between the two platforms. Some of the tables show performance numbers of the abstract reducers for comparison.

Demanding tests based on the pure Lambda Calculus are chosen to make comparisons with the abstract reducers possible and to exercise only the "core" features of these implementations. All the results should be interpreted under this premise.

Before the en masse listing of the tables we make the following comments:

- The first series of tests (Table 7.12) consists of mutual application of Church Numerals; this application amounts to exponentiation and is highly demanding. In these tests applicative and normal order are performing poorly and all three systems face roughly the same limits.

- The *pred** tests (Table 7.13) show the same exponential growth under both applicative and pure normal order evaluation; applicative order is actually quite worse. Lazy evaluation on the other hand, obtained by adding sharing, manages to reduce the complexity to quadratic on the size of the input. *HOR-SH* shows dramatic reductions in reduction counts and time.

- The *fsa** tests (Tables 7.14 and 7.15) belong to a family which is presented by Frandsen and Sturtivant [Frandsen 91] as a proof of the existence of expressions which have an infinitely increasing ratio of minimal vs. "parallel" reduction counts (μ_β/μ_π). This family is defined in Table 7.10 and its second member is shown in tree form in Figure 7.17. All members of this family reduce to $I = (\text{-}1\ 0)$.

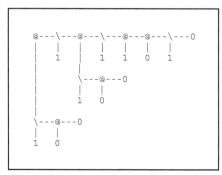

Figure 7.17 Expression fsa2 = ((-1 0 0)(-1 (-1 0 0)(-1 1 (0 (-1 0))))) in tree form.

Test	Definition
fsa1	((-1 0 0) (-1 0 (-1 0)))
fsa2	((-1 0 0) (-1 (-1 0 0) (-1 1 (0 (-1 0)))))
fsa3	((-1 0 0) (-1 (-1 0 0) (-1 (-1 0 0) (-1 2 (1 (0 (-1 0)))))))
fsa4	((-1 0 0) (-1 (-1 0 0) (-1 (-1 0 0) (-1 (-1 0 0) (-1 3 (2 (1 (0 (-1 0))))))))), etc.

Table 7.10 Definition of the fsa* expression family.

- Applicative order and *RTNF/RTLF* with sharing produce identical results on the *fsa** tests . Normal order is hopeless and sharing under *HOR* is not as efficient even though it appears to perform the minimal number of reductions. We have not managed to obtain a result for *fsa4* under *HOR-SH* on our experimental platform.

- Performance of *HORSE* is compromised because of its more complex dispatch mechanism for instructions. The abstract pure HOR reducers show reduction rates that are proportionally higher given the platform performance ratio and the respective coding styles.

- Finally, the tests of Table 7.16 consist of equality checking of Church Numerals entirely within the pure calculus. A *THOR* definition and *HORSE* run appears in Table 7.11. The *cheq** tests show that although *HOR-SH* is initially at a disadvantage it eventually becomes the best strategy. This is not an uncommon situation. The higher overhead of *HOR-SH* in some cases results in a lower degree of complexity and slower overall growth and therefore proves helpful in some large problems.

THOR Definition	HORSE Run
| ```
(define K_
  (lambda (x y) y))

(define succ
  (lambda (n x y) (n x (x y))))

(define pred
  (lambda (n)
    (n (lambda (y) (y n K_ K_ (succ y)))
       (lambda (x y z w) z))))

(define zerot
  (lambda (a b c)
    (a (lambda (d e) e) (lambda (f g) g)
       (lambda (h) c) b)))

(define cheq
  (lambda (a b)
    (rec (lambda (c d e)
           (a d (a e
                   (lambda (f g) f)
                   (lambda (h i) i))
              (a e (lambda (j k) k)
                 (c (b d) (b e)))))))
    zerot
    pred)
``` | ```
horse> c4
(LAMBDA (X Y) (X (X (X (X Y)))))
1 reductions, total.
Elapsed time: 0.000 seconds.

horse> c5
(LAMBDA (X Y) (X (X (X (X (X Y))))))
1 reductions, total.
Elapsed time: 0.000 seconds.

horse> cheq c5 c5
(LAMBDA (F G) F)
22887 reductions, total.
Elapsed time: 0.472 seconds.

horse> cheq c4 c5
(LAMBDA (H I) I)
9481 reductions, total.
Elapsed time: 0.171 seconds.

horse> cheq c5 c4
(LAMBDA (J K) K)
9481 reductions, total.
Elapsed time: 0.172 seconds.
``` |

Table 7.11 Definition and run of cheq, a lambda-defined equality test for Church Numerals.

	π-RED*		π-RED$^+$		HORSE	
	red.†	sec.	red.	sec.	red.	sec.
c2a2	8	0.00	5	0.00‡	6	0.000
c3a3	19	0.00	15	0.00‡	26	0.001
c4a4	92	0.02	87	0.00‡	170	0.017
c5a5	789	0.15	783	0.00‡	1 562	0.199
c5a7	2 808	0.68	2 802	0.00‡	5 602	0.701
c6a6	(1)	—	(1)	—	18 862	3.416
c7a7	(2)	—	(3)	—	(4)	—
NOTES	(1) Failure: "Unexpected descriptor class". (2) Failure: "Out of m stack space". (3) Failure: "Out of heap space, >15M entries allocated". (4) Failure: "Out of control stack space, >400K entries allocated". $\dagger$ Extra applications were added to achieve *nf* in the case of π-RED*. $\ddagger$ Reported "processing" times. There was a post-processing cost of 0.86 and 4.71 secs for *c5a5* and *c5a7* respectively.					

Table 7.12 π-REDs/HORSE. Mutual application of Church Numerals.

	π-RED*		π-RED+		HORSE		HOR§	HOR-SH*
	red.†	sec.	red.	sec.	red.	sec.	sec.‡‡	sec.‡‡
pred2	15	0.00	9	0.00	23	0.001	0.003	0.005
pred3	28	0.01	15	0.00	50	0.001	0.006	0.007
pred4	62	0.01	26	0.00	107	0.003	0.012	0.012
pred5	244	0.01	65	0.01	224	0.007	0.023	0.015
pred6	1 933	0.14	345	0.05	461	0.015	0.044	0.021
pred7	22 964	1.62	3 311	0.52	938	0.027	0.091	0.026
pred8	(1)	—	42 582	8.72††	1 895	0.056	0.182	0.033
pred9			(2)	—	3 812	0.112	0.365	0.042
pred10					7 649	0.224	0.733	0.051
pred11					15 326	0.451	1.465	0.060
pred12					30 683	0.902	5.086	0.071
pred13					61 400	1.804	8.052	0.083
pred14					122 837	3.594	16.084	0.096
pred15					245 714	7.201	(3)	0.110
pred16					491 471	11.307‡		0.125
NOTES	(1) Failure: "Segmentation fault". (2) Failure: "Out of heap space, >15M entries allocated". (3) Stopped due to excessive GC; *pred14* used >3.6M heap space. § Reduction counts for *HOR* are identical to that of *HORSE*. * Reduction counts for *HOR-SH* are given in Table 7.7. † Extra reductions were manually requested to achieve *nf* in the case of π-RED* but the additional time is not included. †† π-RED+ used >14M heap entries. ‡ *HORSE* used >1.5M entries for redex store. ‡‡ *Macintosh Common Lisp*, 25MHz 68030.							

Table 7.13 π-REDs/HORSE. Predecessor function on Church Numerals.

	π-RED*		π-RED+		HORSE	
	red.	sec.	red.	sec.	red.	sec.
fsa1	4	0.00	4	0.00	4	0.000
fsa2	20	0.00	20	0.00	31	0.001
fsa3	184	0.02	184	0.00	(1)	—
fsa4	9 720	0.45	9 720	0.19		
fsa5	20 827 484	1341.6‡	(2)	—		
NOTES	(1) Out of graph memory, >10M entries allocated. (2) Failure: "Invalid CLOS ftype". ‡ Did not attempt to continue with *fsa6*.					

Table 7.14 Family with infinitely increasing μ_β/μ_π of Frandsen & Sturtivant.

	π-RED*		AP-SH		HOR-SH	
	red.	*sec.*	*red.*	*sec.†*	*red.*	*sec.†*
fsa1	4	0.00	4	0.001	4	0.001
fsa2	20	0.00	20	0.003	12	0.004
fsa3	184	0.02	184	0.024	122	0.031
fsa4	9 720	0.45	9 720	1.379	(1)	—
fsa5	20 827 484	1341.6	(1)‡	—		
NOTES	(1) Stopped after 600 secs † *Macintosh Common Lisp, 25MHz 68030.* ‡ Did not attempt to continue with *fsa6*.					

Table 7.15 π-RED*, AP-SH and HOR-SH. Family with infinitely increasing μ_β/μ_π. Comparison of π-RED* with the abstract sharing reducers.

	HORSE		HOR		AP-SH		HOR-SH	
	red.‡	*sec.*	*red.‡*	*sec.†*	*red.*	*sec.†*	*red.*	*sec.†*
cheq0	26	0.000	23	0.002	23	0.003	28	0.031
cheq1	70	0.001	59	0.006	59	0.006	55	0.047
cheq2	198	0.004	159	0.016	113	0.012	96	0.067
cheq3	588	0.011	455	0.045	193	0.021	155	0.092
cheq4	2 488	0.048	1 875	0.189	319	0.036	236	0.127
cheq5	22 888	0.470	17 015	1.720	523	0.059	343	0.173
cheq6	528 656	10.317	391 147	57.536	849	0.098	480	0.233
cheq7	(1)	—	(2)	—	1 353	0.159	651	0.312
cheq8					21 03	0.249	860	0.411
cheq9					3 179	0.380	1 111	0.533
cheq10					4 673	0.562	1 408	0.684
cheq11					6 689	0.803	1 755	0.865
cheq12					9 343	1.127	2 156	1.081
cheq13					12 763	1.542	2 615	1.343
cheq14					17 089	2.071	3 136	1.641
cheq15					22 473	5.002	3 723	1.994
NOTES	† *Macintosh Common Lisp, 25MHz 68030, no instrumentation.* ‡ Differences in reduction counts are due to slightly different definitions. (1) "Out of memory" failure. (2) Did not attempt to continue with *cheq7*.							

Table 7.16 Comparison of HORSE with the abstract reducers on the equality test on Church Numerals.

7.7 REMARKS

Reduction Rates

Reduction rates of about 10K reductions/second are measured with our minimal *Lisp* implementation of *HOR* (without sharing) on a 25MHz 68030[40]. We estimate that with a similar environment on a recent high-performance workstation one can easily achieve 300-500K reductions/second. With sharing, reduction counts can drop dramatically but reduction rates "typically" drop only by factors between two and five. Hence, a conservative estimate for a lazy reduction simulator running on a current technology workstation is in the order of 100K reductions/sec. The reader should keep in mind that such reduction rates refer to a λ-calculus *interpreter* which does not include any preprocessing or post processing, is expression rather than ground value-based, and does error checking and typing at runtime[41]. With low level coding, streamlined representations and perhaps local in-line compilation around strict primitives, one should be able to gain one more order of magnitude. Another order of magnitude speedup should be achievable if a state-of-the-art *HOR*-specific VLSI implementation is attempted. Hence, rates of 1-10M reductions/second are within sight.

We believe that these estimates are not unreasonable and they seem to make compilation techniques—which typically are not expression-based and depend on strong typing—much less attractive. Finally, all these estimates are for sequential implementations. If one manages to take advantage of the parallelism inherently present in the operational head normal forms then there is potential for symbolic computation of staggering performance indeed.

Value of a Reduction

Our comment regarding the "unit value" of accomplishing one reduction being too small to justify elaborate mechanisms for the selection of a redex may be appreciated more when the following empirical information is contemplated.

---

[40] Compare with *HORSE* which on a SUN server 4/490 is typically clocked at rates of up to 30-65K reductions/sec.

[41] Not unlike "logical inference" rates (*LIPS*), reduction rates are not by themselves a very meaningful measure. Reduction of lambda expressions incurs costs even if few reductions are possible. In the extreme, any expression in *nf* will exhibit a null reduction rate but will consume real resources. Reduction rates do provide a measure of maximum raw performance, especially when reported on specific benchmark cases together with absolute reduction counts. One should keep in mind that sharing mechanisms while promoting efficiency actually reduce the observed reduction rates—sometimes dramatically.

	HOR-SH-VA Normal Order Experimental Reducer/Simulator all* test suite, Reds.=769, Calls=21 752 Mem=376K	Total Time† (Secs)	Speedup relative to C.	Slowdown relative to C.
A.	B3-SH, fully instrumented, low-level simulator/reducer with a linear environment.	8.30	0.13	7.61
B.	HOR-SH-VA, instrumented with reduction and iteration counts via three array updates; also full stack frames, tracing conditionals, etc.	3.02	0.36	2.77
C.	HOR-SH-VA, instrumented with reduction and iteration counts via three global variable updates.	1.09	1.00	1.00
D.	HOR-SH-VA, no instrumentation (as in Appendix A).	0.93	1.17	0.85
E.	Like D. with optimizations via the CLtL2 compiler.	0.84	1.30	0.77
F.	Like E. with interrupts disabled.	0.81	1.35	0.74
Note: † All tests performed under Macintosh Common Lisp, 25MHz 68030 processor.				

Table 7.17 Effect of instrumentation on the performance of the experimental reducers.

The augmentation of a minimal abstract reducer like *HOR-SH* with the barest of instrumentation[42] increases runtime by 17%. Also, in our experimental setup, the addition of three updates of a global instrumentation array which collect individual statistics for each test expression nearly triples the observed runtimes[43]. For completeness sake, and to give a better feel for these measurements that support the "unit-value" vs. "unit-cost" conviction, we summarize in Table 7.17 the total time spent on reducing the general test suite *all** under six different simulation situations.

For reference we repeat that plain *HOR* on the same test set performs a total of 13 147 reductions and requires 48 998 iterations. Without any instrumentation and with compiler optimizations turned on it takes 1.81 seconds[44]. Table D.2 lists detailed statistics under *HOR-SH-VA*.

Reduction Counts as Measure of Efficiency

The core assumption underlying the usual optimality criterion is that the number of β-contractions performed in reducing the expression is a good measure of the total costs of normalization. The following quote typifies this assumption.

---

[42] Updating of three global variables that hold counts of the number of reductions and the number of invocations of *up* and *down*.

[43] The high cost is due to the very flexible nature of arrays of *Common Lisp* which necessitates many run-time checks and "out-of-line" code.

[44] Like case *E.* of Table 7.17; i.e., *CL's (declare (optimize (safety 0) (speed 3)))*.

– 159 –

... It seems impossible to find uniform cost measure that works in all situations and thus, we hand-wave a little. First, we assume that any interpreter of the λ-calculus uses an operation that corresponds to reducing a β-redex. Second, we assume that the cost of this operation is about the same as the cost of reducing a β-redex in the usual framework of the λ-calculus. (Note the actual cost of reducing a β-redex is not constant but depends on the redex). Third, we assume that if the interpreter performs other operations, then the cost of these operations can be subsumed in the cost of operation that is analogous to reducing a β-redex. ... [Kathail 90].

But the constant of "proportionality" varies widely, and can be made—by choosing especially contrived cases—arbitrarily large. Hence, the whole premise about proportionality looses its weight. Corroborating arguments are provided by the same author[45].

Nature and Cost of Copying

In proposals centering on optimality one sometimes encounters the situation of contracting redices in isolation. Often it is witnessed while following the trace of environment values and by observing that a redex (often a trivial one, like Ix) has been contracted in one step and hence duplicate reductions are avoided in case it is referred to multiple times subsequently. This, indeed, can be made to happen but not without the cost of determining whether the resulting term is closed so that it may be substituted in parallel in all of its needed contexts. In addition, the practical issues of retrieving an environment expression at the proper time, reducing it in isolation and updating existing structures to reflect the effect of reduction are often downplayed.

Another area which sometimes appears to be confusing is the nature of copying.

... Use of closures obviates copying any part of the body of an abstraction after β-contraction. By contrast, Wadsworth's graph reduction scheme copies the parts of an abstraction containing the abstraction's bound variable, in order to avoid incorrect substitutions in pieces of the abstraction's body that might be shared by other terms. By using environments, the body of the abstraction term, and hence any redexes contained therein, have the potential to be shared, avoiding redundant reductions. ... [Field 91].

Copying cannot be avoided entirely. An expression in head position must be allowed to interact with distinct operands. With delayed substitution copying is piecemeal—a new part of the result expression is constructed by partially replicating and interacting with an

---

[45] A proof of an argument about the unbounded magnitude of such a constant of "proportionality" appears in the appendix of V. Kathail's thesis [Kathail 90].

older portion. Result expressions under graph reduction are acyclic directed graphs (with the cyclic exception of backward pointing variables of REP-style representations) and therefore even if at some intermediate stage an operator is captured by a compact graph structure, it may eventually need to be unraveled in order to produce further results. Hence, an undue emphasis on avoiding copying is limiting and unnatural.

Copying itself is not a simple notion. The very use of the term "copying" implies the creation of a physical structure that unnecessarily duplicates an existing one. Under delayed-substitution techniques copying is manifested by the creation of new portions of the result term based on the "template" of the original. It is only slightly more expensive than a mere traversal which, arguably, has to happen anyway. The significant cost of copying is that it may lead to a loss of sharing and so, indirectly, to duplicate reductions. But performing extra reductions is not so costly. Our belief, supported by empirical observations, is that a reducer quickly reaches the point of diminishing returns; additional bookkeeping efforts required for avoiding contractions are costlier than the potential gains. Nonetheless, further research of a realistic attitude is bound to produce better strategies.

Given suitable representations, the costs of β-reduction—including that of judicious copying—are comparable to the costs expended in combinator graph reduction, a result which should not be surprising given the foundational affinity of Lambda and Combinator Calculi. The reader is reminded of the short discussion of section 1.4.

When a minimal reducer like *HOR* is studied, it becomes clear that any effort expended in minimizing the number of reductions should be compared with and judged against the costs incurred by additional structures and additional rules. One can imagine situations where an interpreter with a certain strategy and representation does particularly well on one or more classes of specially designed test cases but we are convinced that tactics of that sort are not applicable when the goal is to find generally efficient algorithms on which to base the design of a reduction system.

Optimal = Efficient ?

More work is needed to test the hypothesis of whether the quest for optimality can produce methods which are efficient. But most times this hypothesis is taken as fact.

> ... However one is usually not interested in whether an interpreter finds a minimum length reduction or not; one is more interested in the overall efficiency of an interpreter. Thus, an interpreter that matches the efficiency of a minimum length reduction is as good as an interpreter that finds a minimum length reduction. ... [Kathail 90].

Passages like the above illustrate vividly the prevailing psychology with respect to reduction and efficiency. Even while admitting that overall efficiency need not be tied to the length of the reduction one finds that the accorded measure of efficiency remains deeply ingrained. But most of our colleagues are also quick to accept that optimality appears to be an elusive objective and empirical work tends to confirm this suspicion.

> ... On the negative side, there could be an argument on the complexity of the labels. In order to find the correct output edge, it may be necessary to test the equality of the labels which [may] involve a complex algorithm, which could be of a complexity similar to the history of a reduction. For instance, to have the Church Rosser property, it is mandatory to have an associativity property inside the labels and thus really work with character strings. ... [Lévy 88].

Again, by necessity, this goal of unbounded prescience cannot be reached in a practical setting. The Lambda Calculus can model situations of arbitrary complexity and the goal of devising a reducer which is uncommonly cunning about minimizing reductions without penalty would be akin to a global optimization procedure, something that is unlikely to exist.

Chapter 8

Towards Practical Reduction Systems

8.1 ACTIVE VS. PASSIVE GRAPHS

One of the overall goals of this research is to produce a practical reduction system
with a simple core language, normal-order evaluation and a reasonable degree of sharing.
Such a system is based on a conservative extension of the Lambda Calculus. In this chapter
we visit some of the issues raised by the low-level procedural implementations of Head Order
Reduction (HOR) and by the augmentation of the calculus with primitives and data
structures. We also demonstrate the implications for parallel reduction exposed by the
Breadth-First HOR strategy.

Berkling's work has this total focus and applied emphasis since the early seventies
[Berkling 77]. Michael Hilton's *THOR/HORSE* addresses these issues and implements
procedural solutions, albeit under a pure normal-order strategy [Hilton 90b]. The π-*RED*
systems demonstrate that, among other features, an interactive visual environment which
can expand and collapse expressions on demand makes a system with reduction semantics
much more useful [Kluge 93].

The amount of compilation that is allowable in an expression-based system is rather
minimal. Expressions must be easily reconstructible from their machine representations.
This means the abstraction and applicative structure should be preserved and user-level
identifiers must be maintained throughout reduction. Even though these identifiers play no
role in the process of reduction it is critical that they are present for the user's benefit. The

protection mechanism is employed to disambiguate nested occurrences of binders with the same identifier but with a "crossing" binding structure.

Lambda Expressions as Instruction Streams

When the emphasis of the implementation of a reduction mechanism shifts from a passive interpreter of graph expressions to that of an equivalent (hardware or software) automaton the responsibility of control becomes distributed. The state manipulations that are to be effected can be captured by a set of instructions. Berkling has proposed such an organization for the "Lambda Calculus Computer" concept [Berkling 92] and Hilton has refined it further for *THOR/HORSE* [Hilton 90b].

> ... The first step towards developing an architecture is to transform the expression graphs from passive data structures that are interpreted by the *SSM* control unit into active instructions that drive the reduction process. ... The linear graph representation is used as the basis for the *μRED* instruction set. Each graph node becomes an instruction with an opcode and a data field. The node's type becomes the instruction's opcode, and the instruction's data is either a de Bruijn index or a pointer.
> ...[Hilton 90b], p. 45.

For a reduction system based on HOR, such an instruction set has a one-to-one correspondence with the operational structures present in the abstract reducers. For example, an abstraction node becomes a "lambda" (*LAM*) instruction and its primary effect—pushing a closure or a *UVC* onto the environment—is initiated when the instruction is encountered by the control unit. Execution of such a program takes a rather peculiar slant: the instruction representation of the problem expression is "executed" and the effect is that another "program", the normal form, is produced.

Linear Coding of Graphs

Central to the success of this shift from passive interpretation to active machine instructions is a representation of graphs as linear sequences of instructions and an effective organization of memory. Locality of reference is vital in minimizing the creation of unnecessary structures and in maximizing performance.

Given that the natural sequencing of HOR depends on the descent of the left spine of the expression the machine representation that is adopted in the one which makes the chaining of the left spine as direct as possible. Nodes of the expression graph along the left spine are stored in adjacent memory locations. Subexpressions emanating from the left spine are stored in subsequent segments of memory and in reverse order of their appearance on the spine. The applicative structure is maintained via forwarding pointers. This linear coding of the left spine is identical to a *cdr*-coding storage of sequences having elements of fixed size. In this instance one can call it "*car*-coding".

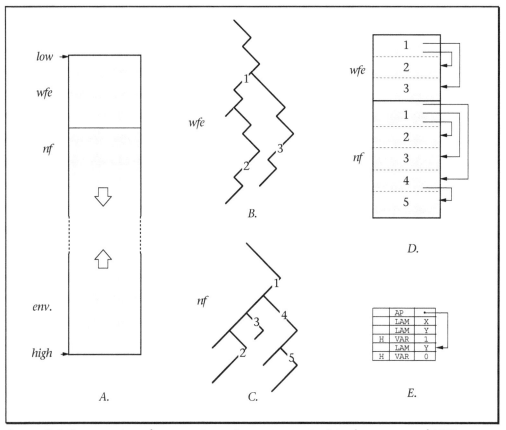

Figure 8.1 Organization of memory in an "active instruction" implementation of HOR.

Figure 8.1 attempts a pictorial view of these decisions. Part (*A*) shows the division of linear memory to a static area allocated to the expression under reduction and a dynamic area that is used during the reduction process. The reduction process is covered in detail in the next section. Parts (*B*) and (*C*) show the problem expression and its normal form as trees with the spines of their subterms emphasized. Part (*D*) shows symbolically how a preorder traversal of these expressions is stored in linear memory. Finally, part (*E*) shows the instruction stream created for *(λxy.x)(λy.y)*. Note that a flag in the leftmost field marks a head variable and hence delineates the corresponding linear segment.

As a more concrete example we show how the term *(λxy.xx)(λuw.uw)(λi.i)* is translated in *THOR/HORSE* to the linear sequence of Figure 8.2. Each instruction occupies two long words or 64 bits. The whole expression requires 96 bytes of memory. Such an allocation is a quite liberal one since it permits 32 bit pointers which can target any location in a 4 gigabyte span of memory.

The first two locations are "apply" or *APP* instructions whose operands are stored in subsequent memory segments. They correspond to the two top-level application nodes of the term. They are coded as such by the use of an '@' flag in the leftmost field of the instruction. This is a useful generalization of the usage of the flag field because it permits an even more economical representation for "short" operands consisting of a single variable or some immediate value.

```
0318:    @ PTR      0368
0320:    @ PTR      0348
0328:      LAMBDA X
0330:      LAMBDA Y
0338:    @ VAR      1
0340:    H VAR      1
0348:      LAMBDA U
0350:      LAMBDA W
0358:    @ VAR      0
0360:    H VAR      1
0368:      LAMBDA I
0370:    H VAR      0
```

Figure 8.2 Linear memory mapping of $(\lambda xy.xx)(\lambda uw.uw)(\lambda i.i)$ in THOR/HORSE.

For comparison, the above expression under the "raw" representation of *BETAR3* requires 64 bytes of memory and under the generic "structural tagging" one employed in our abstract reducers takes up 304 bytes (cf. section 2.2).

Naming Expressions

In a reduction system which allows lambda expressions to be assigned to global names storing of the expressions corresponding to global definitions is a natural extension of the above technique. All the defined expressions reside in a read-only portion of memory. Some means is devised by which the instruction representation for a global name can be located quickly.

The creation of the instruction streams which correspond to symbol definitions and top-level expressions is straightforward; it requires only a depth-first parsing of the input while maintaining a representation environment for identifiers. It is easily performed on-the-fly as an expression is entered in an interactive system or as a file is read in. In the next section we mention some additional processing that can be profitably performed during this phase.

Now we touch upon on the recurring theme of variable representations. The linear coding of graphs makes it clear that applicative structure can be maintained by using relative instead of absolute references. This cannot be easily done in the case of Wadsworth-style pointer variables because the global uniqueness of the corresponding locations is crucial to correctness. Not having to rely on pointers for the representation of the binding structure by using de Bruijn indices instead is a welcome property because simple copying of an expression sequence can replicate its binding and applicative structure without having to maintain a list of binder/target locations.

In summary, the active organization is much more suited to a low-level software or hardware realization. In the next section we present in more detail a fully operational reducer along these lines. What has been accomplished is summed up by Hilton as follows.

> ... Reduction of expressions by the μRED is no longer directed by a monolithic control unit, but is accomplished through the efforts of each instruction comprising the graph. ... This approach streamlines the operation of the μRED control unit to the familiar fetch-decode-execute cycle found in conventional von Neumann architectures. When an instruction is executed, it modifies the machine state according to the appropriate rule. ... [Hilton 90b], p. 49.

Linear Organization of the Environment

The introduction of named expressions in an applied reduction system raises the question of the total organization of the environment.

The side-effect free nature allows us to view this issue from two vantage points. One can consider the process of giving values to symbols which have an associated global definition as an extension of the dynamic environment and incorporate symbol lookup in the general dereferencing procedure. Or one can conceive of symbol definitions as static portions of the top-level program and which are reachable on demand. A newly introduced instruction, "symbol" or *SYM,* which implements the latter option can remain passive when in operand position but it should result to a jump to the associated definition when in head or operator position. If we consider this jump as effecting a reduction then the ordinary mechanism of the remaining contractions quantum can be used to skip the execution of definitions during incremental development. *THOR/HORSE* adopts the latter solution for symbolic constants. Execution of the *SYM* instruction uses a hash-table to retrieve the global definition associated with the symbol.

Of a more difficult nature is the organization of the dynamic portion of the environment. The de Bruijn representation of variables makes lookup a vector indexing operation if the environment is represented as a true vector[46]. But the lexical scoping rules of the calculus result instead in a cactus-like environment structure which in order to be cast into linear memory must be subjected to some form of linking. The *display* mechanism of ALGOL-like languages is not as applicable because of the lack of concrete procedure entry and exit points and the dynamic nature of higher-order functions. We have no magic solution to this problem and we will just remind the reader that in practice one can rely on environment lookups not being very "deep"—once global definitions have been dealt with in the way mentioned previously; therefore, sequential tactics will have to suffice[47].

These ideas and problems are not unique to reduction systems. Similar attitudes towards the organization of environments and similar assessments with regard to the practicality of explicit environments are voiced by other researchers:

> … How efficient can a functional language implementation using the [Categorical Abstract Machine] CAM be? The first problem seems to be access time since we use linked lists for environments instead of arrays. Contrarily to a well agreed opinion, this is not too serious provided that the top-level environment is organized separately. One must keep in mind that in a language like ML the global variables can be dealt with efficiently using a symbol table and the compiler can compute direct access to their value. So the environments which appear in the CAM will only contain free variables of lambda expression or functions which are not global and the length of these environments will thus be limited to the degree of block embedding in a declaration. … On the other hand linked environments enable sharing and fast closure building and this is important when compiling highly functional programs and absolutely necessary if laziness is to be taken into account. … [Cousineau 85].

Recapitulating, the total organization the environment should utilize a hash-table or associative coding for symbol definitions and simple sequential threading for the lexical variables of local lambda definitions. Symbols are active if they are the head of a spine, if they have an associated definition and if the quantum is not exhausted.

---

[46] In the case of entirely dynamic linking for the environment—as in the abstract reducers— lookup is a matter of following exactly n pointers, where n is the de Bruijn index associated to the variable.

[47] Specialized forms of store, like associative or content addressable memory (CAM), can make environment lookups a true constant time operation.

8.2 LOW-LEVEL PROCEDURAL CASTING OF PURE HOR

The operational requirements of HOR are met with relatively simple memory and control structures. Berkling has been focusing on the low-level aspects of reduction ever since a Lambda Calculus machine was conceived. The simplified description of a pure Head Order reducer (μRED) which follows is based on that of Hilton and Berkling [Hilton 90c].

We adopt the organization of linear memory shown in Figure 8.1. The problem expression which is to be reduced is stored at the low end of a segment of memory in the manner discussed in the previous section. This area is accessed in a read mode only—not unlike the program code of the traditional von Neumann machine. The dynamically-sized memory elements of HOR, namely the stack and the environment, share the same memory area and they grow towards each other. Space is available as long as the pointers to the respective endpoints of these areas do not meet. Each location is wide enough to hold two values: a data part (*.d*) and a control part (*.c*).

As the expression graph is traversed its straightened spine is built in the *result* segment of memory and becomes part of the normal form. During the descent of the spine or *forward* traversal the "program counter" points within the expression graph. At the same time whenever records of bindings are made the environment grows towards the result graph. During the ascent of the spine or *reverse* traversal the program counter points within the result graph.

Traversal is organized in such a way that as each *hnf* is computed it is stacked on the result area. *APP* instructions get copied to the result area and remain there unless removed by a subsequent *LAMBDA* instruction. The partial *hnf*s formed in the result area remain undisturbed with the exception of updating the targets of their *APP* instructions to reflect the final location of the reduced operands. This is accomplished via the creation of a *JOIN* instruction before an suspension is reduced. These instructions correspond directly to the marker employed in the abstract HOR family.

A separate LIFO structure, the *control stack* is introduced. As the reduction of each suspension is performed, the environment list grows to include the bindings effected during the reduction of the suspension. The correct initial environment required for the reduction of a suspension is stored in the control stack by pushing the effective environment pointer during the forward execution of an *APP* instruction. In the reverse direction, when reduction of a suspension is completed the environment is cut back by resetting the environment to the value of the pushed pointer.

In the μRED machine eight *registers* suffice; four point to memory and they are best implemented as counters with increment/decrement capability; one is a up/down counter holding negative integer values; one is the single bit direction flag and two are used to store temporary data/address information. Their roles are summarized in Table 8.1.

Reg.	Description
pc	Points to the current instruction to be executed.
sp	Points to the bottom of the result area.
ep	Environment pointer; points to the tip of the current environment.
cp	Control stack pointer; this stack holds the *ep*'s for each operand.
ϕ	The current abstraction-depth context; a negative integer.
d	Direction flag bit; determines *forward* of *backward* execution.
s_a, s_d	Scratch registers used during lookup.

Table 8.1 Description of the registers of the μRED machine.

The initial state of *μRED* consists of a problem expression coded as an instruction stream residing in the expression segment. In this simplified treatment eight *instructions* are enough. Six instructions correspond to the similarly-named components of the abstract HOR reducers. One is the iterative lookup procedure. Finally, a special *STOP* instruction is placed on the top of the result segment to terminate the procedure when the ascent of the normal form is completed.

Figure 8.3 shows all the operational detail. The effect of the *APP* and *LAM* instructions differs depending whether they are encountered in forward or reverse execution. The overall organization is notable for its simplicity. There is also some potential for concurrent operations.

During reduction the result structure and the environment list grow towards each other. When reduction is complete the environment area is empty and the result area holds the final normal form. With the exception of the *JOIN* markers which are present but inactive in the result graph after reduction is complete, the whole procedure does not create any unreachable elements. The normal form is in the same form employed for problem expressions and is ready to participate in further reductions without any post-processing. It is indeed remarkable that a Lambda Calculus reducer—in some respects a superset of a higher-order functional language—can be operationally so well behaved.

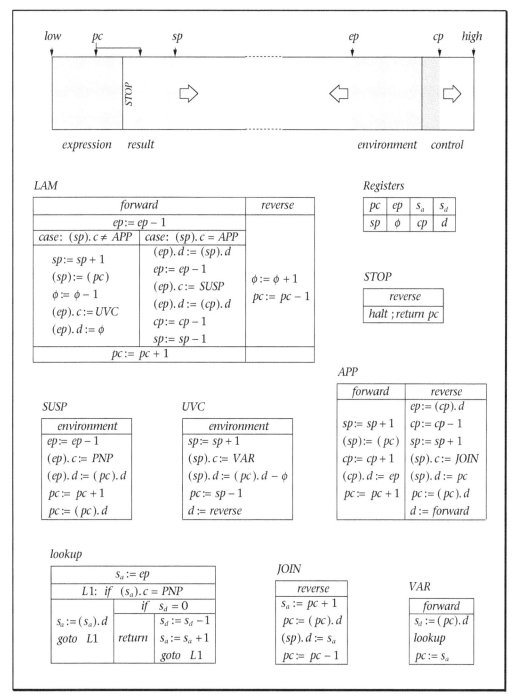

Figure 8.3 The μ-RED low-level pure HOR machine with linear-memory.

The previous low-level casting of a pure, non-sharing version of HOR is an excellent fit to linear memory. An equivalent operational casting for an optimized version of *HOR-SH* is more complicated and it does not share such an ideal fit. This is because the sharing versions of HOR do require recursive calls of the overall strategy during reduction of an expression in isolation. The problems are envisioned mostly in the organization of the environment area.

8.3 ENRICHED CALCULUS

The Lambda Calculus is capable of expressing and simulating all possible computations. A reduction system however does not begin to match the utility of a general purpose computing environment unless it is enriched with problem-oriented elements. Ideally, the addition of such elements must be in accordance with the underlying theoretical model in the sense that functionality afforded in the extended domain would have been possible in the domain of the pure calculus.

It is outside the scope of this dissertation to attempt to provide concrete suggestions for the multitude of the programming constructs and structuring mechanisms that have been devised in decades of computer science research. This is a substantial undertaking on its own with many exciting possibilities. But it also includes the danger of entering interminable arguments like the ones of computer language design. Infinite variations are possible; the right mix of features ultimately depends on intended usage. Simplicity and general applicability should be the guiding principles.

In the following we sketch how some of the most common programming constructs can be emulated. We start be reminding the reader of the Lambda Calculus definitions of a some basic constructs (Table 8.2). The first two show suitable encodings for the truth values. The last one shows that structures can be built by making their constituent elements operands of a head normal form.

Lambda Definition	Operation
$T \equiv \lambda xy.x$	*True*
$F \equiv \lambda xy.y$	*False*
$if\ C\ \ then\ \ E_t\ \ else\ \ E_f \equiv CE_tE_f$	*Conditional*
$f \circ g \equiv \lambda x.f(gx)$	*Composition*
$[A,B] \equiv \lambda h.hAB$	*Pairing*

Table 8.2 Lambda Calculus definitions of common operations.

A practical reduction system should provide constants for numbers, characters, and other commonly used atomic elements. These elements are passive in the sense of reducing to themselves. The next step is to add the related primitives which combine and test the

3-list in *HORSE*	
| [A, B, C]<br><br>$(\lambda.p\,(p\;A)$<br>$\qquad(\lambda.p\,(p\;B)$<br>$\qquad\qquad(\lambda.p\,(p\;C)\;NIL)))$ | ```
9a0:  P  LAMBDA $$P
9a8:  @  PTR     9c0
9b0:  @  SYMBOL  A
9b8:  H  VAR     0
9c0:  P  LAMBDA $$P
9c8:  @  PTR     9e0
9d0:  @  SYMBOL  B
9d8:  H  VAR     0
9e0:  P  LAMBDA $$P
9e8:  @  SYMBOL  NIL
9f0:  @  SYMBOL  C
9f8:  H  VAR     0
``` |

Table 8.3 Lists can be encoded as nested head normal forms.

values of such constants like arithmetic operators, relational tests and type predicates. These primitives are naturally strict; i.e., they force the reduction of their operands. In a completely general system the representation of constants typically includes a tag signifying their type so that primitive operators can decide whether they are to fire.

Table 8.3 shows how a list can be profitably coded as a head normal forms (*hnf*). Projection is effected by applying the structure to a selection combinator like those shown in Table 1.5. A selection combinator is a closed lambda expression of the form (– *m n*) which has to be *applied* to such structures. After one reduction the combinator is substituted for the head variable of the *hnf* and therefore has access to all arguments subterms. The overhead of such coding is not high and access of a component is implemented efficiently via indexing into the environment. This technique is easily extended to nested structures with corresponding multi-level selectors. It is the lambda calculus equivalent of array indexing and it is possible to implement it similarly.

Firing of Primitives

In a system in which the evaluation order is generally non-strict the implementation of strict primitives depends on reducing the proper number of arguments of such primitives. This can be accomplished by keeping track of the number of active arguments in the current *hnf*. In the BTRD experimentation system this mechanism is included so that the behavior of combinator reduction can be emulated. A combinator or a primitive in head position does not "fire" unless the requested number of arguments are present in the spine. In *HORSE* the mechanism for primitives consists of a register, *prim*, that holds the code of the last primitive encountered in operator position and a counter, *fire*, which is initially loaded with the required number of arguments of the primitive in *prim*. When reduction of an operand is
```

completed the *fire* counter is decremented. When it reaches zero the proper number of arguments have been reduced and the primitive is allowed to fire.

Since the arguments of a strict primitive can be arbitrary expressions which can themselves include other strict primitives the values of these two registers have to stacked each time the reduction of a new argument is initiated. The values are restored when reduction of an argument is completed. In a sequential software simulation these requirements present an appreciable overhead to what is otherwise a unit-time operation. In a hardware implementation the testing of the registers and the pushing and popping of their values can be done in parallel with other processing.

If a primitive finds that its arguments do not have the proper type it does not signal an error; simply the primitive does not "fire". Firing of primitives results in instant reclamation of space since the result overwrites the operator and the arguments.

In *THOR/HORSE* structures are made passive when they are not in operator position i.e., when are not applied to any arguments. The reduction quantum is temporarily zeroed during traversal which effectively implements lazy constructors.

Inclusion of a predicate for equality is a complicated issue. Equality is not decidable within the Lambda Calculus because of the non-normalizable terms. In practice there is no difficulty of providing a test for syntactic structural equivalence if an equality test is made strict in both arguments. The name-free representations result in identical sequences for α-convertible terms; therefore the equality test is simply a traversal in tandem.

In-Line Compilation

Arithmetic and logic expressions consisting solely of variables and strict primitives do not require the full generality of the graph representation. Such expressions can be expressed via simpler code that uses the result area as a pushdown stack.

Table 8.4 shows that an arithmetic expression can be compiled into a flat linear representation which is concise and suitable for direct execution by a stack-based automaton. The applicative structure is replaced by a sequencing in Reverse Polish Notation which is suitable for in-line execution.

Recursion

There are many ways to implement recursion. The traditional method "native" to the calculus is via the *Y* fixed-point combinator which has the property of replicating its function argument. Typically, a special rewrite rule is used for *Y* in order to achieve more efficient processing but to also to make the presence of recursion easy to spot. Other possibilities commonly present in programming languages are recursion via *global* symbol definitions and via the creation of a *letrec* extended context.

In *HORSE* recursion via a global definitions relies on the "symbol" (*SYM*) instruction. Table 8.5 shows how the tree of a definition of the factorial function is mapped into linear memory. When traversal reaches location *c60* forward execution continues by reentering the definition of *fact*.

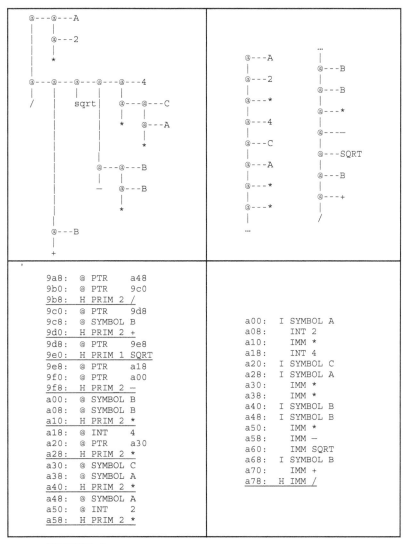

```
@---@---A
|   |
|   @---2
|   |
|   *
|
@---@---@---@---@---4
|   |   |   |   |
/   |  sqrt|   @---@---C
    |      |   |   |
    |      |   *   @---A
    |      |       |
    |      |       *
    |      |
    |      @---@---B
    |      |   |
    |      |   -   @---B
    |      |       |
    |      |       *
    |
    @---B
    |
    +
```

```
@---A                       ...
|                           |
@---2                       @---B
|                           |
@---*                       @---B
|                           |
@---4                       @---*
|                           |
@---C                       @----
|                           |
@---A                       @---SQRT
|                           |
@---*                       @---B
|                           |
@---*                       @---+
|                           |
...                         /
```

```
9a8:   @ PTR     a48
9b0:   @ PTR     9c0
9b8:   H PRIM 2  /
9c0:   @ PTR     9d8
9c8:   @ SYMBOL B
9d0:   H PRIM 2  +
9d8:   @ PTR     9e8
9e0:   H PRIM 1  SQRT
9e8:   @ PTR     a18
9f0:   @ PTR     a00
9f8:   H PRIM 2  −
a00:   @ SYMBOL B
a08:   @ SYMBOL B
a10:   H PRIM 2  *
a18:   @ INT     4
a20:   @ PTR     a30
a28:   H PRIM 2  *
a30:   @ SYMBOL C
a38:   @ SYMBOL A
a40:   H PRIM 2  *
a48:   @ SYMBOL A
a50:   @ INT     2
a58:   H PRIM 2  *
```

```
a00:   I SYMBOL A
a08:   INT 2
a10:   IMM *
a18:   INT 4
a20:   I SYMBOL C
a28:   I SYMBOL A
a30:   IMM *
a38:   IMM *
a40:   I SYMBOL B
a48:   I SYMBOL B
a50:   IMM *
a58:   IMM −
a60:   IMM SQRT
a68:   I SYMBOL B
a70:   IMM +
a78:   H IMM /
```

Table 8.4 $(b + \sqrt{b^2 - 4ac})/2a$ under generic and in-line instruction streams.

The results of a short experiment with the different possibilities of providing recursion is shown in Tables 8.6 and 8.7. One observes that under the pure normal order sequencing of *HORSE* all four methods have comparable runtime behavior with the possible exception of environment space. Note that the pure normal order implies that the runtime requirements of these tests have a non-linear component (cf. section 7.5).

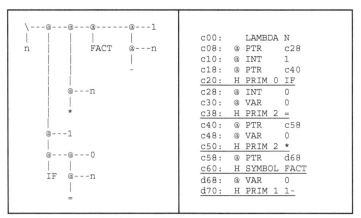

```
\---@---@---@------@---1                 c00:      LAMBDA N
 |   |   |   |      |                     c08:  @ PTR     c28
 n   |   |  FACT   @---n                  c10:  @ INT     1
     |   |          |                     c18:  @ PTR     c40
     |   |          -                     c20:  H PRIM 0 IF
     |   |                                c28:  @ INT     0
     |  @---n                             c30:  @ VAR     0
     |   |                                c38:  H PRIM 2 =
     |   *                                c40:  @ PTR     c58
     |                                    c48:  @ VAR     0
    @---1                                 c50:  H PRIM 2 *
     |                                    c58:  @ PTR     d68
    @---@---0                             c60:  H SYMBOL FACT
     |   |                                d68:  @ VAR     0
    IF  @---n                             d70:  H PRIM 1 1-
         |
         =
```

Table 8.5 Tree and linear graph of fact $\equiv \lambda n.\text{if} (= n\ 0)\ 1\ (*\ n\ (\text{fact}\ (- n\ 1))))$.

```
;;; Global symbol                        ;;; LETREC abstraction mechanism
(define gfact
  (lambda (n)                            (define lfact
    (if (= n 0) 1                          (letrec
              (* n (gfact (1- n))))))        ((fact
                                               (lambda (n)
;;; User-defined Y combinator                  (if (= n 0) 1
(define yfact                                        (* n (fact (1- n)))))))
  (Y (lambda (fact n)                        fact))
     (if (= n 0) 1
               (* n (fact (1- n)))))))    ;;; The Y combinator

;;; System-defined Y combinator          (define Y
(define rfact                              (lambda (x)
  (rec (lambda (fact n)                      ((lambda (y) (y y))
       (if (= n 0) 1                          (lambda (z) (x ( z z)))))))
               (* n (fact (1- n)))))))
```

Table 8.6 THOR definitions of four versions of the factorial function.

Program	Reductions	Max Graph	Max Env.	Max Stack	Instructions	Time (secs)
gfact	$6n+4$	4 010	4 001	3 007	24 015	.542
rfact	$7n+6$	4 010	8 005	3 007	28 018	.637
yfact	$7n+9$	4 010	10 015	3 007	33 028	.723
lfact	$6n+5$	4 010	4 006	3 007	25 019	.562

Table 8.7 HORSE statistics of fact 3000 under the four definitions of factorial.

This inefficiency leads us again to sharing. As we have seen effective sharing in the presence of recursion is an absolute requirement for a practical lambda reduction system. An eloquent exposition of this point has also been made by Kieburtz [Kieburtz 86]. We have seen how pure *HOR* suffers from an unacceptable increase in the normalization effort due to normal order. The sharing version of *BETAR3* solves this problem partially by implementing a cyclic version of the *Y* when sharing and the rewrite rule *Yf → f(Yf)* are used in conjunction. Under sharing a heap must be used for storing structures resulting from reduction in isolation since it cannot be easily anticipated if they are going to be needed again.

Finally, one should not fail to make use of the information that can be gathered easily the first time an expression is typed or read-in. During read-in expressions can be compiled in-line if they consist solely of primitives and also if they have a flat binding level. Subterms that are already in *nf* can be tagged as such. Common subexpressions can also be identified during read-in and the result of this analysis can be reflected in the graph representation. At runtime detecting whether an expressions is closed is possible with little additional effort. Such tagging can save unneeded corrective traversals for closed expressions which were formed under a different abstraction context.

8.4 IMPLICATIONS OF BF-HOR FOR PARALLEL GRAPH REDUCTION

Opportunities for parallelism abound during reduction of lambda expressions. By projection the same holds for applied systems based on the Lambda Calculus. At the lowest level the natural intermediate stage of reduction under HOR, the operational *hnf*, exposes the work that may beneficially be done in parallel.

The properties of the breadth-first reducer (section 4.4) that are central to the goal of exploring parallelism are summarized below.

- By definition, the *hnf* has the same overall structure as the *nf*.

- Reduction of the arguments (*susps* in *ohnf*) could be done in parallel with no interaction.

- All suspensions originating from the same *hnf* share the *hnf's* abstraction-depth context (ϕ tag).

- A suspension emanating from an *ohnf* needs (read) access only to the environment that was significant during the formation of the *ohnf*.

- After reduction of each suspension is complete, the environment portion created during such reduction is not needed anymore and can be safely discarded.

Under this light, *BF-HOR* reduction appears to have important implications for close-knit automatic parallel reduction. It exposes the inherent parallelism present in an expression via the process queue. Each element of the queue can be processed independently of all others and in parallel with other elements.

Our hope is that relatively simple heuristics can be devised to control such parallelism. For example, it may make sense to continue all the work on the same processor unless the formation of a new *ohnf* has increased the number of elements of the process queue by a certain number or by a certain percentage. The fact that the base of the environment—i.e., the portion which is common among the suspensions emanating from the same *hnf*—is shared and read-only is most helpful.

Figure 8.4 shows graphically some of these interesting aspects of *BF-HOR* reduction. A profile of the parallelism is displayed for five expressions. Each row corresponds to one down-up traversal of a delayed suspension and includes three pieces of information. In the first column a large dot signals a reduction. In the second column the process queue is shown by utilizing a bar for each element. The third column is an estimate of the effort of the down-up traversal for the last suspension processed. In this example the effort is measured in $(ms{\cdot}bytes)^{1/2}$.

One observes that there is no assurance that the elements of the queue pack comparable amounts of work. In some cases—see for instance the profile of *pred3*—a process queue never actually builds up. All the effort of normalization is expended in the reduction of the first delayed suspension. In other cases like *tt23* the process queue consists of a single element throughout while reductions are distributed; there is no opportunity to divide the queue. We also see that the *VA* optimization can have a pronounced effect on both the average depth of the queue and on the number of trivial suspensions (observe the two traces given for each of *tw7-2* and *(tw10 tw7)*). As expected, *VA* reduces substantially the number of trivial suspensions that pack no work other than looking up a variable.

We suspect that without source transformations which explicitly invite parallelism signatures like that of *tt23* or *pred3* would be a common theme among many programming tasks. It remains to be seen how sharing affects this aspect of breadth-first reduction.

One should keep in mind that parallelism of this sort is primarily theoretically possible—not unlike the *AND-OR* parallelism that surfaces when refutation of a goal clause is attempted in a Horn-clause logic program. Inventing general mechanisms which manage to consistently extract run-time benefits has proved difficult.

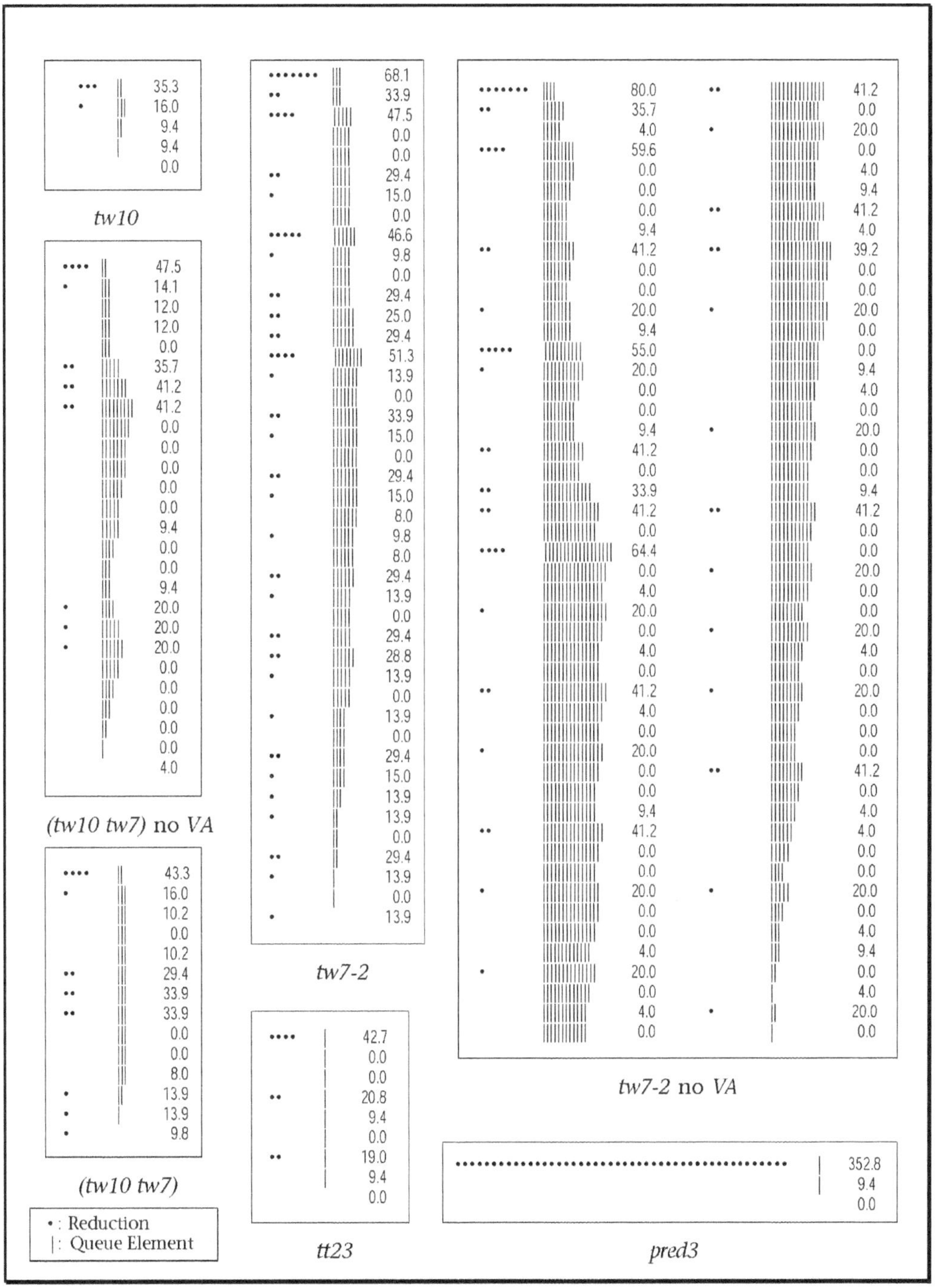

Figure 8.4 Process queues under **BF-HOR** for five test expressions together with the number of reductions and the cost of each traversal.

Causal interdependencies and parallelism

In the field of compilation-centered parallel graph reduction it is not uncommon to encounter the proposal that a strict separation of reduction and parallelism can lead to competent parallel implementations. This view is often taken due to the fact that sequential implementations have reached high-levels of performance and therefore the thinking is that one should be able to drop freely a specific sequential implementation as a component in a parallel system without special regard on the details of the specific sequential reducer.

In light of the concepts and the techniques we have described, we believe that parallelism is an inherent property of the overall normalization process and in particular of the *hnf*s and ideally should be exploited as such. It is only then that one can ensure that the unavoidable efforts required when multiple agents collaborate is justifiable in light of the overall goal of producing a normal form. This fact has resulted in statements like the following by proponents of the above mentioned separation [Langendoen 93].

> "To catch up [with] the performance [of] non-functional competitors like object-oriented based systems, many researchers follow the advice of [Vrancken90] and concentrate on advancing sequential compilation technology. This trend is likely to change future parallel implementations of functional languages since fast implementations are rather sensitive to runtime support overhead costs."

The separation can be applied only when parallelism is exploited at the rather coarse level of named function calls; a by-product of compilation-based transformations which do not necessarily expose the inherent causal interdependencies and parallelism present in a program as a whole.

8.5 FINAL REMARKS

Future Work

- Feasibility of sharing in a *HOR* reducer but with a pointer variable representation.

- Exploration of options for adding sharing under Breadth-First *HOR*.

- Addition of various forms of sharing under Head-Normal *HOR*.

- Augmentation of the sharing mechanism of *HOR* with versions that rely on weak and head-normalization of the operator expressions.

- Applicability of the techniques of unbound variable counts and context tagging in other domains.

- Extension of the focus domain to an applied-calculus with a *Y*-rule, syntactic equality predicate, delta primitives and arithmetic optimizations in the presence of sharing. Exploration of the repercussions of these additions to the basic mechanisms of *BF-HOR* and *HOR-SH*.

- Low-level casting for *HOR* with sharing. Emphasis on the organization of the environment. Extension to a fully applied calculus.

- Further study of the issues raised by cyclic data structures.

- Development of a machine/chip architecture with the above applied-calculus as its native instruction set.

- Exploration of ways of characterizing and controlling the inherent parallelism exposed by Breadth-First HOR. Interaction with cyclic/infinite data structures.

- Application work in the areas of parallel data structures, experimentation with the semantics for process Calculi.

Conclusions-Summary

We highlight below some of the insights gained from the development of the techniques that we have described in this thesis.

- Two major ingredients are required in order to make strongly-normalizing reduction systems with normal-order strategy practical. First, in order to behave acceptably in terms of time and space one must be able to effectively share intermediate results of expressions during the reduction process. The second one is the ability to operate declaratively— using the native machinery of the Lambda Calculus if possible—on cyclic implementations of data structures while avoiding exponential explosion. In this dissertation *HOR-SH* and its variations provide new options with respect to the former and it may be the case that Breadth-First *HOR* constitutes progress towards the latter.

- A β-contraction does not involve a lot of work. In delayed substitution reducers, one only needs to make a record of the association between binder and substituted expression, a very simple operation. Strong normalization is not expensive with the introduction of the *UVC* mechanism. Updating de Bruijn indices as the result of delayed substitutions is simply a matter of applying a correction when an index is encountered during traversal. A correction has only has to be done if a variable occurrence is part of the final result. The formulation of the rules for HOR in Table 3.8 corresponding to the abstract rewrite equations (3.7) define the core of what we believe is the minimal machinery for efficient reduction.

- References to the use of closures as a delaying mechanism for achieving β-contractions are often clouded by an erroneous view. The comparison is made with Wadsworth's reducer need to copy the portions of an abstraction's body—which can be a shared operator of other applications—that contain the abstraction's bound variable and the implication is made that environments and closures can somehow avoid this copying. We argued that even those reducers which delay substitutions have to effectively copy the abstraction's body. Copying from an abstraction's body that was encountered in operator position cannot be avoided because the body may need to interact with distinct operands. The important question is not how to avoid copying but how to do it efficiently and as part of the total effort of constructing a normal form. Reduction in isolation of a suspension in operator position appears to provide substantial benefits because it reduces the frequency of such copying and forces it on generally shorter terms.

- Sharing is absolutely vital in conjunction with normal order reduction. Normal order with sharing betters applicative order not only in terms of termination but also economy. There are varying degrees of laziness as John Field has argued, from zero (normal order's complete lack of sharing), to Lévy's theoretical optimality. Aiello–Prini's *AP-SH* and our *HOR-SH* reducers appear to strike operationally feasible and practically interesting balances between economy and optimality. We showed that the *HOR-SH* tactic is unsafe because an out-of-context reduction of a diverging but solvable term may fail to terminate. *HOR-SH* is incomplete but to a lesser degree than eager evaluation. At the same time it is more economical.

- A breadth-first sequencing has been presented which postpones reduction of arguments as long as desired. Under a *BF-HOR* strategy, the suspensions pending from each traversed spine are added to a global process queue. Each element of the queue consists of an expression-environment pair, an integer signifying the abstraction depth context of the parent *hnf* and a pointer recording the location of origin on the spine. This information is necessary and sufficient for each suspension to be lazily reduced in isolation. After a suspension is itself transformed to *hnf* the application node of the parent spine is modified to point to the new structure. *BF-HOR* exposes the inherent parallelism present in the evaluation of a lambda expression. It also introduces the possibility of sharing via the process queue.

- Graph reduction does not have to rely on pointers for the representation of bindings. Parallel implementations of graph reduction can benefit from representations which do not rely on pointer variables because the extra costs of maintaining the binding structure at this minute level when graphs are moved to independently-addressed memory spaces can be avoided. Graph representations which are not readily relocatable are not well suited to dynamic systems.

- A common assumption is that the mere presence of an environment structure makes a reducer less efficient. An environment is an orderly manner of keeping track of bindings, a requirement that has to fulfilled somehow in a dynamic system. With indexing based on de Bruijn variables a lookup is nearly a constant time operation. An environment does have the benefit of a less ad hoc memory allocation and deallocation and the possibility of automatic memory reclaim. In the sharing versions of *HOR,* usage of environment space is extremely modest. The breadth-first versions of *HOR* do suffer from higher environment residency because the reduction of the pending closures is piecemeal and no definite return point exists where a linear environment can be trimmed.

- We refrained from delving into comparisons between direct lambda calculus reduction and combinator techniques because combinators are inferior both in terms of maintaining transparency and intent throughout reduction and in terms of efficiency. This is so because the difficulties attributed to relatively free variables can be overcome when de Bruijn indices and the *UVC* mechanism for closing expressions artificially is introduced.

- The low-level procedural casting of HOR demonstrates that it is possible to implement a pure higher-order functional language with reduction semantics without garbage collection. The reason behind this is the well regimented memory requirements of the HOR scheme. All the activity happens at the tip of the straightened spine which coincides with the location where the normal form is built. The situation is not as fortunate though when sharing is introduced.

The picture of direct lambda reduction is forming as follows. A Lambda Calculus reducer must make choices for all of the following: (*a*) overall strategy for choosing redices, which is related intrinsically to the order of traversal; (*b*) immediate or delayed substitutions; (*c*) representations for variables and open-terms; (*d*) implementation of sharing, i.e., how are the terms to be shared chosen and how is sharing effected; (*e*) early or late variable argument lookups, i.e., the extent of the "eagerness" of argument suspension dereferencing, and (*f*) other detailed representation choices like the realization of the binding and applicative structure.

The overall strategy and order of traversal is the most important decision but it is via all the above elements that a reducer becomes practical. Our experimental system currently includes reducers which encompass more than thirty variations of the previous choices.

In terms of strategy we believe that the virtues of proceeding to the head along the extended left spine were made clear. Breadth-first reduction of operands is promising. Delayed substitutions are definitely preferable to literal ones. Early variable argument dereferencing avoids excessive laziness. Reducing subterms from operator position and then selectively copying from the ensuing normal forms offers the most pronounced gains.

A reducer should must make good choices for all of the above issues. But no single reducer seems to win on all fronts. Finally, even though no formal arguments were made we believe we demonstrated that direct lambda reduction is preferable to abstraction techniques—even in terms of economy. *BTF* by itself can be considered a form of combinatory logic corresponding directly to lambda reduction.

The problems of effective structure sharing and infinite (cyclic) data structures are among the oldest in implementations of declarative computing. One case in point is logic programming research which has devoted a lot of effort in devising scalable methods that guide operational behavior without affecting declarative meaning. This ideal is not always reached and compromises are often made. It could well be that techniques like *HOR-SH* and, eventually, parallel versions of *BF-HOR* with sharing fall under the same category of imperfect but practically interesting proposals.

Appendices

A. ABSTRACT REDUCERS IN NEARLY PURE LISP

In this appendix we list the concrete *Lisp* definitions of representative reducers discussed throughout this thesis. The reader is referred to Table 1.6 for an overview, classification and symbolic names for these reducers.

All the abstract reducers operate on representations of lambda expressions described in section 2.2. A pure functional style is employed throughout. The only non-pure constructs are the in-situ replacements effected, solely in the event of sharing, to structures pointed by environment elements. They are implicit in the macros *susp2clos*, *susp2shexp* and the use of *setf* in *AP-HOR*.

```
(defun eval (term env)

  (if (is-var term)
    (val-of (look-up term env))

    (if (is-app term)
      (apply (eval (fun-of term) env) (eval (arg-of term) env))

      (if (is-lam term)
        (mk-clos term env)

        (if (is-clos term)
          (term-of term))))))

(defun apply (rator rand)

  (let ((fun (term-of rator))
        (env (env-of rator)))
    (eval (form-of fun) (mk-bind (bnd-of fun) rand env))))

(defun look-up (var env)
  (if (eq var (var-of env))
    env
    (look-up var (rest-of env)))))
```

Figure A.1 (EV-AP). McCarthy's Eval-Apply as Lambda Calculus Reducer.

```
(defun secd (stack env cntl dump)

  (if (null cntl)
    (if (null dump)
      (top-of stack)
      (secd
       (push (top-of stack) (dump-s dump))
       (dump-e dump)
       (dump-c dump)
       (dump-d dump)))

   (let ((toc (pop cntl)))

     (if (is-var toc)
       (secd (push (nth toc env) stack) env cntl dump)

       (if (is-marker toc)
         (let ((clos (pop stack)))
           (secd
            nil
            (mk-hor-bind (pop stack) (env-of clos))
            (list (frm-of (term-of clos)))
            (mk-dump stack env cntl  dump)))

         (if (is-lam toc)
           (secd (push (mk-clos toc env) stack) env cntl dump)

           (if (is-app toc)
             (secd
              stack
              env
              (push-3 (arg-of toc) (fun-of toc) marker cntl)
              dump)))))))))
```

Figure A.2 (SECD-V). Landin's SECD Applicative Order Reducer.

```
(defun hor (term env phi)

  (if (is-hor-var term)
    (let ((val (nth term env)))
      (if (is-hor-susp val)
        (hor (term-of val) (env-of val) phi)
        (- val phi)))

    (if (is-app term)
      (if (is-lam (fun-of term))
        (hor (frm-of (fun-of term)) (mk-hor-bind (mk-susp (arg-of term) env) env) phi)

        (let ((rator (hor (fun-of term) env phi)))
          (if (is-lam rator)
            (hor (frm-of rator) (mk-hor-bind (mk-susp (arg-of term) env) (mk-dummy-env phi)) phi)
            (mk-app rator (hor (arg-of term) env phi)))))

      (if (is-lam term)
        (mk-lam (mk-var '\_) (hor (frm-of term) (mk-hor-bind (1- phi) env) (1- phi)))))))
```

Figure A.3 (HOR-NS) Normal-Form Reducer without an explicit stack.

```
(defun rtnf (term lexenv)

  (if (is-var term)
    (let ((env (look-up term lexenv)))
      (let ((susp (val-of env)))
        (if (is-susp susp)
          (rtnf (term-of susp) (env-of susp))
          susp)))

    (if (is-app term)
      (let ((susp (rtlf (fun-of term) lexenv)))
        (if (is-susp susp)
          (let ((fun (term-of susp))
                (env (env-of susp)))
            (rtnf (form-of fun)
              (mk-bind (bnd-of fun)
                    (mk-susp (arg-of term) lexenv)
                    env)))
          (mk-app susp (rtnf (arg-of term) lexenv))))

      (if (is-lam term)
        (let ((newvar (mk-var (nam-of (bnd-of term)))))
          (mk-lam newvar
            (rtnf (frm-of term)
              (mk-bind (bnd-of term) newvar lexenv)))))))))


(defun rtlf (term lexenv)

  (if (is-var term)
    (let ((env (look-up term lexenv)))
      (let ((susp (val-of env)))
        (if (is-susp susp)
          (rtlf (term-of susp) (env-of susp))
          susp)))

    (if (is-app term)
      (let ((susp (rtlf (fun-of term) lexenv)))
        (if (is-susp susp)
          (let ((fun (term-of susp))
                (env (env-of susp)))
            (rtlf (form-of fun)
              (mk-bind (bnd-of fun)
                    (mk-susp (arg-of term) lexenv)
                    env)))
          (mk-app susp (rtnf (arg-of term) lexenv))))

      (if (is-lam term)
        (mk-susp term lexenv)))))


(defun look-up (var env)
  (if (or (eq var (var-of env)) (null env))
    env
    (look-up var (rest-of env))))
```

Figure A.4 (AP) Aiello and Prini's RTNF/RTLF Reducer.

```
(defun rtnlf (term env f)

  (if (is-var term)
    (let* ((env (look-up term env))
           (susp (val-of env)))
      (if (is-susp susp)
        (let ((val (rtnlf (term-of susp) (env-of susp) f)))
          (if SH (setf (val-of env) val) val))
        (if (and (is-lam susp) (not f)) (mk-susp susp nil) susp)))

    (if (is-app term)
      (let ((susp (rtnlf (fun-of term) env nil)))
        (if (is-susp susp)
          (if (and VA (is-var (arg-of term)))
            (rtnlf (frm-of (term-of susp))
                   (mk-bind (bnd-of (term-of susp))
                            (val-of (look-up (arg-of term) env)) (env-of susp)) f)
            (rtnlf (frm-of (term-of susp))
                   (mk-bind (bnd-of (term-of susp))
                            (mk-susp (arg-of term) env) (env-of susp)) f))
          (mk-app susp (rtnlf (arg-of term) env t))))

      (if f
        (let ((newvar (mk-var (gensym))))
          (mk-lam newvar (rtnlf (frm-of term) (mk-bind (bnd-of term) newvar env) t)))
        (mk-susp term env)))))
```

Figure A.5 (AP-SH-VA) RTNF/RTLF Reducer with sharing, early VA and force flags.

```
(defun whor (term stack env)

  (if (is-hor-var term)
    (let ((val (nth term env)))
      (whor (term-of val) stack (env-of val)))

    (if (is-app term)
      (if (is-hor-var (arg-of term))
        (whor (fun-of term) (push (nth (arg-of term) env) stack) env)
        (whor (fun-of term) (push (mk-susp (arg-of term) env) stack) env))

      (if (is-lam term)
        (let ((tos (top-of stack)))
          (if (is-clos tos)
            (whor (frm-of term) (pop-stack stack) (mk-hor-bind tos env))
            (mk-clos term env)))))))
```

Figure A.6 (W-HOR-VA) Weak HOR Reducer.

```
(defun aphor (term env phi f)

  (if (is-hor-var term)
    (let ((val (nth term env)))
      (if (is-hor-susp val)
        (let ((rterm (aphor (term-of val) (env-of val) phi f)))
          (if SH
            (if (is-susp rterm)
              (setf (nth term env) rterm)
              (susp2shexp (nth term env) rterm phi)))
            rterm)
        (if (is-hor-uvc val)
          (- val phi)
          (if (is-sh-exp val)
            (let ((sphi (phi-of-sh-exp val))
                  (sterm (term-of-sh-exp val)))
              (if (= sphi phi)
                (if (or f (not (is-lam sterm))) sterm (mk-susp sterm (mk-arid)))
                (aphor sterm (mk-dummy-env sphi) phi f)))))))

  (if (is-app term)
    (let ((susp (aphor (fun-of term) env phi nil)))
      (if (is-hor-susp susp)
        (if (and VA (is-hor-var (arg-of term)))
          (aphor (frm-of (term-of susp))
                 (mk-hor-bind (nth (arg-of term) env) (env-of susp)) phi f)
          (aphor (frm-of (term-of susp))
                 (mk-hor-bind (mk-susp (arg-of term) env) (env-of susp)) phi f))
        (mk-app susp (aphor (arg-of term) env phi t))))

  (if (is-lam term)
    (if f
      (mk-lam (mk-var '\_) (aphor (frm-of term) (mk-hor-bind (1- phi) env) (1- phi) t))
      (mk-susp term env))))))
```

Figure A.7 (AP-HOR) Reducer with RTNF/RTLF strategy and dB/UVC machinery including sharing and early VA.

```
(defun whorsh (term stack env)

  (if (is-hor-var term)
    (let ((val (nth term env)))
      (if (is-hor-susp val)
        (susp2clos val (whor (term-of val) nil (env-of val))))
      (whorsh (term-of val) stack (env-of val)))

  (if (is-app term)
    (if (is-hor-var (arg-of term))
      (whorsh (fun-of term) (push (nth (arg-of term) env) stack) env)
      (whorsh (fun-of term) (push (mk-susp (arg-of term) env) stack) env))

  (if (is-lam term)
    (let ((tos (top-of stack)))
      (if (or (is-clos tos) (is-hor-susp tos))
        (whorsh (frm-of term) (pop-stack stack) (mk-hor-bind tos env))
        (mk-clos term env)))))))
```

Figure A.8 (W-HOR-SH-VA) Sharing version of Weak HOR Reducer.

```
(defun down (term stack env phi)

 (if (is-hor-var term)
   (let ((val (hor-look-up term env)))
     (if (is-hor-uvc val)
       (up (- val phi) stack phi)
       (down (term-of val) stack (env-of val) phi)))

   (if (is-app term)
     (if (is-hor-var (arg-of term))
       (down (fun-of term)(push (hor-look-up (arg-of term) env) stack) env phi)
       (down (fun-of term)(push (mk-susp (arg-of term) env) stack) env phi))

     (if (is-lam term)
       (let ((tos (top-of stack)))
         (if (or (is-hor-susp tos) (is-hor-uvc tos))
           (down (frm-of term) (pop-stack stack)(mk-hor-bind tos env) phi))
           (down (frm-of term) (push *lambda-marker* stack)
                 (mk-hor-bind (1- phi) env) (1- phi)))))))))


(defun up (rterm stack phi)

 (if (null stack)
   rterm

   (let ((tos (pop stack)))

     (if (is-join-marker tos)
       (up (mk-app (pop stack) rterm) stack phi)

       (if (is-hor-susp tos)
         (down (term-of tos) (push-2 *join-marker* rterm stack) (env-of tos) phi)

         (if (is-lambda-marker tos)
           (up (mk-lam (mk-var '\_) rterm) stack (1+ phi))

           (if (is-hor-uvc tos)
             (up (mk-app rterm (- tos phi)) stack phi)))))))))
```

Figure A.9 (HOR-VA) Berkling's Head Order Reducer with Early VA lookup.

```
(defun rdown (term stack env)

  (if (is-var term)
    (let* ((renv (look-up term env))
           (susp (val-of renv)))
      (if (is-susp susp)
        (rdown (term-of susp) stack (env-of susp))
        (rup susp stack)))

    (if (is-app term)
      (if (and VA (is-var (arg-of term)))
        (rdown (fun-of term) (push (val-of (look-up (arg-of term) env)) stack) env)
        (rdown (fun-of term) (push (mk-susp (arg-of term) env) stack) env))

      (if (is-lam term)
        (let ((tos (top-of stack)))
          (if (or (is-var tos) (is-susp tos))
            (rdown (frm-of term) (pop-stack stack) (mk-bind (bnd-of term) tos env))
            (let ((newvar (mk-var (mk-var '\_))))
              (rdown (frm-of term) (push (cons newvar *lambda-marker* ) stack)
                     (mk-bind (bnd-of term) newvar env)))))))))


(defun rup (rterm stack)

  (if (null stack)
    rterm

    (let ((tos (pop stack)))

      (if (is-join-marker tos)
        (rup (mk-app (pop stack) rterm) stack)

        (if (is-susp tos)
          (rdown (term-of tos) (push-2 *join-marker* rterm stack) (env-of tos))

          (if (is-lambda-marker (cdr tos))
            (rup (mk-lam (car tos)  rterm) stack)

            (if (is-var tos)
              (rup (mk-app rterm tos) stack)))))))))
```

Figure A.10 (HOR-AP) Head Order Reducer with REP-style structures.

```
(defun down (term stack env phi)

  (if (is-hor-var term)
    (let ((val (hor-look-up term env)))
      (if (is-hor-uvc val)
        (up (- val phi) stack phi)
          (if (is-sh-exp val)
            (if (and (or (is-join-marker (top-of stack)) (null stack)) (= (phi-of-sh-exp val) phi) )
              (up (term-of-sh-exp val) stack phi)
              (down (term-of-sh-exp val) stack (mk-dummy-env (phi-of-sh-exp val)) phi))
            (progn
              (susp2shexp val (down (term-of val) nil (env-of val) phi) phi)
              (if (or (is-join-marker (top-of stack)) (null stack))
                (up (term-of-sh-exp val) stack phi)
                (down (term-of-sh-exp val) stack (mk-dummy-env (phi-of-sh-exp val)) phi))))))

  (if (is-app term)
    (if (is-hor-var (arg-of term))
      (down (fun-of term) (push (hor-look-up (arg-of term) env) stack) env phi)
      (down (fun-of term) (push (mk-susp (arg-of term) env) stack) env phi))

  (if (is-lam term)
    (let ((tos (top-of stack)))
      (if (or (is-hor-susp tos) (is-hor-uvc tos) (is-sh-exp tos))
        (down (frm-of term) (pop-stack stack) (mk-hor-bind tos env) phi)
        (down (frm-of term) (push *lambda-marker* stack)
          (mk-hor-bind (1- phi) env) (1- phi)))))))))


(defun up (rterm stack phi)

  (if (null stack)
    rterm

    (let ((tos (pop stack)))

      (if (is-join-marker tos)
        (up (mk-app (pop stack) rterm) stack phi)

        (if (is-hor-susp tos)
          (down (term-of tos) (push-2 *join-marker* rterm stack) (env-of tos) phi)

          (if (is-lambda-marker tos)
            (up (mk-lam (mk-var '\_) rterm) stack (1+ phi))

            (if (is-hor-uvc tos)
              (up (mk-app rterm (- tos phi)) stack phi)

              (if (is-sh-exp tos)
                (if (= (phi-of-sh-exp tos) phi)
                  (up (mk-app rterm (term-of-sh-exp tos)) stack phi)
                  (down (term-of-sh-exp tos) (push-2 *join-marker* rterm stack)
                    (mk-dummy-env (phi-of-sh-exp tos)) phi)))))))))))
```

Figure A.11 (HOR-SH-VA) Head-Order Reducer with sharing, early VA lookup and optimizations.

```
(defun bfh (term)
  (let ((app (mk-app nil (mk-susp term (mk-arid)))))
    (arg-of (bfhlist app (list (mk-prs app 0))))))


(defun bfhlist (hnfterm pqueue)
  (if (null pqueue) hnfterm
    (let* ((app  (app-of (top-of pqueue)))
           (phi  (phi-of (top-of pqueue)))
           (susp (arg-of app))
           (skeleton  (bfhdown (term-of susp) nil (env-of susp) phi (rest-of pqueue))))
      (setf (arg-of app) (hnfterm-of skeleton))
      (bfhlist hnfterm (pqueue-of skeleton)))))


(defun bfhdown (term stack env phi pqueue)

  (if (is-hor-var term)
    (let ((val (nth term env)))
      (if (is-hor-susp val)
        (bfhdown (term-of val) stack (env-of val) phi pqueue)
        (bfhup (- val phi) stack phi pqueue)))

    (if (is-app term)
      (if (is-hor-var (arg-of term))
        (bfhdown (fun-of term) (push (nth (arg-of term) env) stack) env phi pqueue)
        (bfhdown (fun-of term) (push (mk-susp (arg-of term) env) stack) env phi pqueue))

      (if (is-lam term)
        (let ((tos (top-of stack)))
          (if (or (is-hor-var tos) (is-hor-susp tos))
            (bfhdown (frm-of term) (pop-stack stack) (mk-hor-bind tos env) phi pqueue)
            (bfhdown (frm-of term) (push lambda-marker stack)
              (mk-hor-bind (1- phi) env) (1- phi) pqueue)))))))


(defun bfhup (rterm stack phi pqueue)

  (if (null stack)
    (list rterm pqueue)

    (let ((tos (pop stack)))

      (if (is-hor-susp tos)
        (let ((app (mk-app rterm tos)))
          (bfhup app stack phi (nconc pqueue (list (mk-prs app phi)))))

        (if (is-lambda-marker tos)
          (bfhup (mk-lam (mk-var '\_) rterm) stack (1+ phi) pqueue)

          (if (is-hor-var tos)
            (bfhup (mk-app rterm (- tos phi)) stack phi pqueue)))))))
```

Figure A.12 (BF-HOR) Breadth-First HOR Reducer.

```
SH
(progn
  (susp2shexp val (down (term-of val) nil (env-of val) phi) phi)
  (down (term-of-sh-exp val) stack (mk-dummy-env (phi-of-sh-exp val)) phi))

VA
(if (is-hor-var (arg-of term))
  (down (fun-of term) (push (hor-look-up (arg-of term) env) stack) env phi)
  (down (fun-of term) (push (mk-susp (arg-of term) env) stack) env phi))

NEW
(if (or (is-join-marker (top-of stack)) (null stack))
  (up (term-of-sh-exp val) stack phi)
  (down (term-of-sh-exp val) stack (mk-dummy-env (phi-of-sh-exp val)) phi))

READY
(if (and (or (is-join-marker (top-of stack)) (null stack)) (= (phi-of-sh-exp val) phi) )
  (up (term-of-sh-exp val) stack phi)
  (down (term-of-sh-exp val) stack (mk-dummy-env (phi-of-sh-exp val)) phi))

SH.SUSP.UP
(progn
  (susp2shexp tos (down (term-of tos) nil (env-of tos) phi) phi)
  (if (or (is-join-marker (top-of stack)) (null stack))
    (up (mk-app rterm (term-of-sh-exp tos)) stack phi)
    (down (term-of-sh-exp tos) (push-2 *join-marker* rterm stack)
        (mk-dummy-env (phi-of-sh-exp tos)) phi)))

SHEXP.UP
(progn
  (susp2shexp tos (down (term-of-sh-exp tos) nil (mk-dummy-env (phi-of-sh-exp tos)) phi) phi)
  (up (mk-app rterm (term-of-sh-exp tos)) stack phi))
```

Figure A.13 Code fragments defining the variations of HOR.

B. REDUCTION TRACES/GRAPHS

This appendix displays full reduction traces and other detailed information that was mentioned throughout this thesis and which is too unwieldy to be included in place.

	pred2	whnf	hnf, nf
0.	((-1 0 (-1 0 1 (-2 0) (-2 0) (-2 2 1 (1 0))) (-4 1)) (-2 1 (1 0)))		
1.	(((-2 1 (1 0)) (-1 0 (-2 1 (1 0)) (-2 0) (-2 0) (-2 2 1 (1 0))) (-4 1))		
2.	((-1 0 (-2 1 (1 0)) (-2 0) (-2 0) (-2 2 1 (1 0)) ((-1 0 (-2 1 (1 0)) (-2 0) (-2 0) (-2 2 1 (1 0))) 0)) (-4 1))		
3.	((-1 0 (-2 1 (1 0)) (-2 0) (-2 0) (-2 2 1 (1 0))) ((-1 0 (-2 1 (1 0)) (-2 0) (-2 0) (-2 2 1 (1 0))) (-4 1))		
4.	((-1 0 (-2 1 (1 0)) (-2 0) (-2 0) (-2 2 1 (1 0)) (-4 1)) (-2 1 (1 0)) (-2 0) (-2 0) (-2 (-1 0 (-2 1 (1 0)) (-2 0) (-2 0) (-2 2 1 (1 0)) (-4 1) 1 (1 0)))		
5.	((-4 1) (-2 1 (1 0)) (-2 0) (-2 (-4 1) 1 (1 0)) (-2 1 (1 0)) (-2 0) (-2 0) (-2 (-1 0 (-2 1 (1 0)) (-2 0) (-2 0) (-2 2 1 (1 0)) (-4 1) 1 (1 0))) 2 1 (1 0))) (-4 1) 1 (1 0)))		
6.	((-3 1) (-2 0) (-2 (-2 (-4 1) 1 (1 0)) (-2 1 (1 0)) (-2 0) (-2 0) (-2 (-2 (-1 0 (-2 1 (1 0)) (-2 0) (-2 0) (-2 2 1 (1 0)) (-4 1) 1 (1 0)) 0)))		
7.	((-2 1) (-2 0) (-2 (-4 1) 1 (1 0)) (-2 1 (1 0)) (-2 0) (-2 0) (-2 (-2 (-1 0 (-2 1 (1 0)) (-2 0) (-2 0) (-2 2 1 (1 0)) (-4 1) 1 (1 0)))		
8.	((-3 0) (-2 (-4 1) 1 (1 0)) (-2 1 (1 0)) (-2 0) (-2 (-2 (-1 0 (-2 1 (1 0)) (-2 0) (-2 0) (-2 2 1 (1 0)) (-4 1) 1 (1 0)))		
9.	((-2 0) (-2 2 1 (1 0)) (-2 0) (-2 (-2 (-1 0 (-2 1 (1 0)) (-2 0) (-2 0) (-2 2 1 (1 0)) (-4 1) 1 (1 0)))		
10.	((-1 0) (-2 0) (-2 0) (-2 (-1 0 (-2 1 (1 0)) (-2 0) (-2 0) (-2 2 1 (1 0)) (-4 1) 1 (1 0)))		
11.	((-2 0) (-2 0) (-2 (-1 0 (-2 1 (1 0)) (-2 0) (-2 0) (-2 2 1 (1 0)) (-4 1) 1 (1 0)))		
12.	((-1 0) (-2 (-1 0 (-2 1 (1 0)) (-2 0) (-2 0) (-2 2 1 (1 0)) (-4 1) 1 (1 0))) (-4 1) 1 (1 0))		
13.	(-2 (-1 0 (-2 1 (1 0)) (-2 0) (-2 0) (-2 2 1 (1 0)) (-4 1) 1 (1 0))		whnf
14.	(-2 (-4 1) (-2 1 (1 0)) (-2 0) (-2 (-4 1) 1 (1 0)) 1 (1 0))		
15.	(-2 (-3 1) (-2 0) (-2 (-4 1) 1 (1 0)) 1 (1 0))		
16.	(-2 (-2 1) (-2 0) (-2 (-4 1) 1 (1 0)) 1 (1 0))		
17.	(-2 (-3 0) (-2 (-4 1) 1 (1 0)) 1 (1 0))		
18.	(-2 (-2 0) 1 (1 0))		
19.	(-2 (-1 0) (1 0))		
20.	(-2 1 0)		hnf, nf

Table B.1 Annotated reduction sequence for pred2.

step	term/shexp	stack	env	phi
D0	(-4 (-3 2 1 0) 3 2)	nil	nil	0
D1	(-3 (-3 2 1 0) 3 2)	()	(-1)	-1
D2	(-2 (-3 2 1 0) 3 2)	(\ \)	(-2 -1)	-2
D3	(-1 (-3 2 1 0) 3 2)	(\ \ \)	(-3 -2 -1)	-3
D4	((-3 2 1 0) 3 2)	(\ \ \ \)	(-4 -3 -2 -1)	4
D5	((-3 2 1 0) 3)	(-2 \ \ \ \)	(-4 -3 -2 -1)	4
D6	(-3 2 1 0)	(-1 -2 \ \ \ \)	(-4 -3 -2 -1)	4
•D7	(-2 2 1 0)	(-2 \ \ \ \)	(-1 -4 -3 -2 -1)	4
•D8	(-1 2 1 0)	(\ \ \ \)	(-2 -1 -4 -3 -2 -1)	4
D9	(2 1 0)	(\ \ \ \ \)	(-5 -2 -1 -4 -3 -2 -1)	-5
D10	(2 1)	(-5 \ \ \ \ \)	(-5 -2 -1 -4 -3 -2 -1)	-5
D11	2	(-2 -5 \ \ \ \ \)	(-5 -2 -1 -4 -3 -2 -1)	-5
U12	4	(-2 -5 \ \ \ \ \)		-5
U13	(4 3)	(-5 \ \ \ \ \)		-5
U14	(4 3 0)	(\ \ \ \ \)		-5
U15	(-1 4 3 0)	(\ \ \ \)		4
U16	(-2 4 3 0)	(\ \ \)		-3
U17	(-3 4 3 0)	(\ \)		-2
U18	(-4 4 3 0)	()		-1
U19	(-5 4 3 0)	nil		0

Table B.2 Test expression #1 under HOR-VA; also under HOR-SH-VA.

step	term	stack	env
D0	((-1 0 0) ((-2 1 1) (-1 0) (-1 0)))	nil	nil
D1	(-1 0 0)	([((-2 1 1) (-1 0) (-1 0)) { }])	nil
•D2	(0 0)	nil	([((-2 1 1) (-1 0) (-1 0)) { }])
D3	0	([((-2 1 1) (-1 0) (-1 0)) { }])	([((-2 1 1) (-1 0) (-1 0)) { }])
D4	((-2 1 1) (-1 0) (-1 0))	([((-2 1 1) (-1 0) (-1 0)) { }])	nil
D5	((-2 1 1) (-1 0))	([(-1 0) { }] [((-2 1 1) (-1 0) (-1 0)) { }])	nil
D6	(-2 1 1)	([(-1 0) { }] [(-1 0) { }] [((-2 1 1) (-1 0) (-1 0)) { }])	nil
•D7	(-1 1 1)	([(-1 0) { }] [((-2 1 1) (-1 0) (-1 0)) { }])	([(-1 0) { }])
•D8	(1 1)	([((-2 1 1) (-1 0) (-1 0)) { }])	([(-1 0) { }] [(-1 0) { }])
D9	1	([(-1 0) { }] [((-2 1 1) (-1 0) (-1 0)) { }])	([(-1 0) { }] [(-1 0) { }])
D10	(-1 0)	([(-1 0) { }] [((-2 1 1) (-1 0) (-1 0)) { }])	nil
•D11	0	([((-2 1 1) (-1 0) (-1 0)) { }])	([(-1 0) { }])
D12	(-1 0)	([((-2 1 1) (-1 0) (-1 0)) { }])	nil
•D13	0	nil	([((-2 1 1) (-1 0) (-1 0)) { }])
D14	((-2 1 1) (-1 0) (-1 0))	nil	nil
D15	((-2 1 1) (-1 0))	([(-1 0) { }])	nil
D16	(-2 1 1)	([(-1 0) { }] [(-1 0) { }])	nil
•D17	(-1 1 1)	([(-1 0) { }])	([(-1 0) { }])
•D18	(1 1)	nil	([(-1 0) { }] [(-1 0) { }])
D19	1	([(-1 0) { }])	([(-1 0) { }] [(-1 0) { }])
D20	(-1 0)	([(-1 0) { }])	nil
•D21	0	nil	([(-1 0) { }])
D22	(-1 0)	nil	nil

Table B.3 W-HOR-VA reduction of ((-1 0 0)((-2 1 1)(-1 0)(-1 0))).

step	term/shexp	stack	env	phi
D0	(-4 (-3 2 1 0) 3 2)	nil	nil	0
D1	(-3 (-3 2 1 0) 3 2)	(\)	(-1)	-1
D2	(-2 (-3 2 1 0) 3 2)	(\ \)	(-2 -1)	-2
D3	(-1 (-3 2 1 0) 3 2)	(\ \ \)	(-3 -2 -1)	3
D4	((-3 2 1 0) 3 2)	(\ \ \ \)	(-4 -3 -2 -1)	4
D5	((-3 2 1 0) 3)	([2 { -4 -3 -2 -1 }] \ \ \ \)	(-4 -3 -2 -1)	4
D6	(-3 2 1 0)	([3 { -4 -3 -2 -1 }] [2 { 4 -3 -2 -1 }] \ \ \ \)	(-4 -3 -2 -1)	4
•D7	(-2 2 1 0)	([2 { -4 -3 -2 -1 }] \ \ \ \)	([3 { -4 -3 -2 -1 }] -4 -3 -2 -1)	4
•D8	(-1 2 1 0)	(\ \ \ \)	([2 { -4 -3 -2 -1 }] [3 { -4 -3 -2 -1 }] -4 -3 -2 -1)	4
D9	(2 1 0)	(\ \ \ \ \)	(-5 [2 { -4 -3 -2 -1 }] [3 { -4 -3 -2 -1 }] -4 -3 -2 -1)	5
D10	(2 1)	([0 { -5 [2 { -4 -3 -2 -1 }] [3 { -4 -3 -2 -1 }] -4 -3 -2 -1 }] \ \ \ \ \)	(-5 [2 { -4 -3 -2 -1 }] [3 { -4 -3 -2 -1 }] -4 -3 -2 -1)	5
D11	2	([1 { -5 [2 { -4 -3 -2 -1 }] [3 { -4 -3 -2 -1 }] -4 -3 -2 -1 }] [0 { -5 [2 { -4 -3 -2 -1 }] [3 { -4 -3 -2 -1 }] -4 -3 -2 -1 }] \ \ \ \ \)	(-5 [2 { -4 -3 -2 -1 }] [3 { -4 -3 -2 -1 }] -4 -3 -2 -1)	5
D12	3	nil	(-4 -3 -2 -1)	-5
U13	4	nil		5
R12	(5.0 . 4)	.	.	
D14	4	([1 { -5 [2 { -4 -3 -2 -1 }] (5.0 . 4) -4 -3 -2 -1 }] [0 { -5 [2 { -4 -3 -2 -1 }] (5.0 . 4) -4 -3 -2 -1 }] \ \ \ \ \)	(-5 -4 -3 -2 -1)	-5
U15	4	([1 { -5 [2 { -4 -3 -2 -1 }] (5.0 . 4) -4 -3 -2 -1 }] [0 { -5 [2 { -4 -3 -2 -1 }] (5.0 . 4) -4 -3 -2 -1 }] \ \ \ \ \)		5
D16	1	(@ 4 [0 { -5 [2 { -4 -3 -2 -1 }] (5.0 . 4) -4 -3 -2 -1 }] \ \ \ \ \)	(-5 [2 { -4 -3 -2 -1 }] (5.0 . 4) -4 -3 -2 -1)	5
D17	2	nil	(-4 -3 -2 -1)	-5
U18	3	nil		5
R17	(5.0 . 3)	.	.	
U19	3	(@ 4 [0 { -5 (5.0 . 3) (5.0 . 4) -4 -3 -2 -1 }] \ \ \ \ \)		-5
U20	(4 3)	([0 { -5 (5.0 . 3) (5.0 . 4) -4 -3 -2 -1 }] \ \ \ \ \)		-5
D21	0	(@ (4 3) \ \ \ \ \)	(-5 (5.0 . 3) (5.0 . 4) -4 -3 -2 -1)	-5
U22	0	(@ (4 3) \ \ \ \ \)		5
U23	(4 3 0)	(\ \ \ \ \)		5
U24	(-1 4 3 0)	(\ \ \ \)		4
U25	(-2 4 3 0)	(\ \ \)		3
U26	(-3 4 3 0)	(\ \)		2
U27	(-4 4 3 0)	(\)		-1
U28	(-5 4 3 0)	nil		0

Table B.4 Test expression #1 under HOR-SH reduction without Early VA.

step	term/clos	env
N0	(-2 (-1 0 (0 2)) ((-2 0 1) 0))	nil
N1	(-1 (-1 0 (0 2)) ((-2 0 1) 0))	(g1)
N2	((-1 0 (0 2)) ((-2 0 1) 0))	(g2 g1)
L3	(-1 0 (0 2))	(g2 g1)
R3	[(-1 0 (0 2)) { g2 g1 }]	
• N4	(0 (0 2))	([[((-2 0 1) 0) { g2 g1 }] g2 g1)
L5	0	([[((-2 0 1) 0) { g2 g1 }] g2 g1)
L6	((-2 0 1) 0)	(g2 g1)
L7	(-2 0 1)	(g2 g1)
R7	[(-2 0 1) { g2 g1 }]	
• L8	(-1 0 1)	([0 { g2 g1 }] g2 g1)
R8	[(-1 0 1) { [0 { g2 g1 }] g2 g1 }]	
R6	[(-1 0 1) { [0 { g2 g1 }] g2 g1 }]	
R5	[(-1 0 1) { [0 { g2 g1 }] g2 g1 }]	
• N9	(0 1)	([(0 2) { [((-2 0 1) 0) { g2 g1 }] g2 g1 }] [0 { g2 g1 }] g2 g1)
L10	0	([(0 2) { [((-2 0 1) 0) { g2 g1 }] g2 g1 }] [0 { g2 g1 }] g2 g1)
L11	(0 2)	([((-2 0 1) 0) { g2 g1 }] g2 g1)
L12	0	([((-2 0 1) 0) { g2 g1 }] g2 g1)
L13	((-2 0 1) 0)	(g2 g1)
L14	(-2 0 1)	(g2 g1)
R14	[(-2 0 1) { g2 g1 }]	
• L15	(-1 0 1)	([0 { g2 g1 }] g2 g1)
R15	[(-1 0 1) { [0 { g2 g1 }] g2 g1 }]	
R13	[(-1 0 1) { [0 { g2 g1 }] g2 g1 }]	
R12	[(-1 0 1) { [0 { g2 g1 }] g2 g1 }]	
• L16	(0 1)	([2 { [((-2 0 1) 0) { g2 g1 }] g2 g1 }] [0 { g2 g1 }] g2 g1)
L17	0	([2 { [((-2 0 1) 0) { g2 g1 }] g2 g1 }] [0 { g2 g1 }] g2 g1)
L18	2	([((-2 0 1) 0) { g2 g1 }] g2 g1)
R18	g1	
R17	g1	
N19	1	([2 { [((-2 0 1) 0) { g2 g1 }] g2 g1 }] [0 { g2 g1 }] g2 g1)
N20	0	(g2 g1)
R20	g2	
R19	g2	
R16	(g1 g2)	
R11	(g1 g2)	
R10	(g1 g2)	
N21	1	([(0 2) { [((-2 0 1) 0) { g2 g1 }] g2 g1 }] [0 { g2 g1 }] g2 g1)
N22	0	(g2 g1)
R22	g2	
R21	g2	
R9	(g1 g2 g2)	
R4	(g1 g2 g2)	
R2	(g1 g2 g2)	
R1	(-1 g1 0 0)	
R0	(-2 1 0 0)	

Table B.5 AP reduction of wads45 = (-2 (-1 0 (0 2))((-2 0 1) 0)) .

step	term / shexp	stack	env	phi
D0	(-2 (-1 0 (0 2)) ((-2 0 1) 0))	nil	nil	0
D1	(-1 (-1 0 (0 2)) ((-2 0 1) 0))	()	(-1)	-1
D2	((-1 0 (0 2)) ((-2 0 1) 0))	(\\)	(-2 -1)	-2
D3	(-1 0 (0 2))	([((-2 0 1) 0) {-2 -1}] \\)	(-2 -1)	-2
•D4	(0 (0 2))	(\\)	([((-2 0 1) 0) {-2 -1}] -2 -1)	-2
D5	0	([(0 2) {[((-2 0 1) 0) {-2 -1}] -2 -1}] \\)	([((-2 0 1) 0) {-2 -1}] -2 -1)	-2
D6	((-2 0 1) 0)	nil	(-2 -1)	-2
D7	(-2 0 1)	([0 {-2 -1}])	(-2 -1)	-2
•D8	(-1 0 1)	nil	([0 {-2 -1}] -2 -1)	-2
D9	(0 1)	()	(-3 [0 {-2 -1}] -2 -1)	3
D10	0	([1 {-3 [0 {-2 -1}] -2 -1}] \)	(-3 [0 {-2 -1}] -2 -1)	3
U11	0	([1 {-3 [0 {-2 -1}] -2 -1}] \)		3
D12	1	(@ 0 \)	(-3 [0 {-2 -1}] -2 -1)	3
D13	0	nil	(-2 -1)	3
U14	1	nil		3
R13	(3.0 . 1)	.	.	
U15	1	(@ 0 \)		3
U16	(0 1)	()		3
U17	(-1 0 1)	nil		-2
R6	(2.0 -1 0 1)		.	
D18	(-1 0 1)	([[(0 2) {(2.0 -1 0 1) -2 -1}] \\)	(-2 -1)	-2
•D19	(0 1)	(\\)	([(0 2) {(2.0 -1 0 1) -2 -1}] -2 -1)	-2
D20	0	([1 {[(0 2) {(2.0 -1 0 1) -2 -1}] -2 -1}] \\)	([(0 2) {(2.0 -1 0 1) -2 -1}] -2 -1)	-2
D21	(0 2)	nil	((2.0 -1 0 1) -2 -1)	-2
D22	0	([2 {(2.0 -1 0 1) -2 -1}])	((2.0 -1 0 1) -2 -1)	-2
D23	(-1 0 1)	([2 {(2.0 -1 0 1) -2 -1}])	(-2 -1)	-2
•D24	(0 1)	nil	([2 {(2.0 -1 0 1) -2 -1}] -2 -1)	-2
D25	0	([1 {[2 {(2.0 -1 0 1) -2 -1}] -2 -1}])	([2 {(2.0 -1 0 1) -2 -1}] -2 -1)	-2
D26	2	nil	((2.0 -1 0 1) -2 -1)	-2
U27	1	nil		-2
R26	(2.0 . 1)	.	.	
D28	1	([1 {(2.0 . 1) -2 -1}])	(-2 -1)	-2
U29	1	([1 {(2.0 . 1) -2 -1}])		-2
D30	1	(@ 1)	((2.0 . 1) -2 -1)	-2
U31	0	(@ 1)		-2
U32	(1 0)	nil		-2
R21	(2.0 1 0)	.	.	
D33	(1 0)	([1 {(2.0 1 0) -2 -1}] \\)	(-2 -1)	-2
D34	1	([0 {-2 -1}] [1 {(2.0 1 0) -2 -1}] \\)	(-2 -1)	-2
U35	1	([0 {-2 -1}] [1 {(2.0 1 0) -2 -1}] \\)		-2
D36	0	(@ 1 [1 {(2.0 1 0) -2 -1}] \\)	(-2 -1)	-2
D37	0	(@ 1 [1 {(2.0 1 0) -2 -1}] \\)		-2
U38	(1 0)	([1 {(2.0 1 0) -2 -1}] \\)		-2
D39	1	(@ (1 0) \\)	((2.0 1 0) -2 -1)	-2
U40	0	(@ (1 0) \\)		-2
U41	(1 0 0)	(\\)		-2
U42	(-1 1 0 0)	()		-1
U43	(-2 1 0 0)	nil		0

Table B.6 HOR -SH (no VA) reduction of wads45 = (-2 (-1 0 (0 2))((-2 0 1) 0)).

step	term/susp	env
N0	(-2 (-1 0 2 0) (-1 0 ((-1 0) 1)))	nil
N1	(-1 (-1 0 2 0) (-1 0 ((-1 0) 1)))	(g1)
N2	((-1 0 2 0) (-1 0 ((-1 0) 1)))	(g2 g1)
L3	(-1 0 2 0)	(g2 g1)
R3	[(-1 0 2 0) { g2 g1 }]	
•N4	(0 2 0)	([[(-1 0 ((-1 0) 1)) { g2 g1 }] g2 g1)
L5	(0 2)	([[(-1 0 ((-1 0) 1)) { g2 g1 }] g2 g1)
L6	0	([[(-1 0 ((-1 0) 1)) { g2 g1 }] g2 g1)
·L7	(-1 0 ((-1 0) 1))	(g2 g1)
R7	[(-1 0 ((-1 0) 1)) { g2 g1 }]	
R6	[(-1 0 ((-1 0) 1)) { g2 g1 }]	
•L8	(0 ((-1 0) 1))	(g1 g2 g1)
L9	0	(g1 g2 g1)
R9	g1	
N10	((-1 0) 1)	(g1 g2 g1)
L11	(-1 0)	(g1 g2 g1)
R11	[(-1 0) { g1 g2 g1 }]	
•N12	0	(g2 g1 g2 g1)
R12	g2	
R10	g2	
R8	(g1 g2)	
R5	(g1 g2)	
N13	0	([[(-1 0 ((-1 0) 1)) { g2 g1 }] g2 g1)
·N14	(-1 0 ((-1 0) 1))	(g2 g1)
N15	(0 ((-1 0) 1))	(g3 g2 g1)
L16	0	(g3 g2 g1)
R16	g3	
N17	((-1 0) 1)	(g3 g2 g1)
L18	(-1 0)	(g3 g2 g1)
R18	[(-1 0) { g3 g2 g1 }]	
•N19	0	(g2 g3 g2 g1)
R19	g2	
R17	g2	
R15	(g3 g2)	
R14	(-1 0 g2)	
R13	(-1 0 g2)	
R4	(g1 g2 (-1 0 g2))	
R2	(g1 g2 (-1 0 g2))	
R1	(-1 g1 0 (-1 0 1))	
R0	(-2 1 0 (-1 0 1))	

Table B.7 AP-SH-VA reduction of wads34 = (-2 ((-1 0 2 0)(-1 0 ((-1 0) 1)))) .

step	term/rator graphs	env/rand graphs
E0	(app (app (lam #1 (lam ◊ (app #1# #1#))) (lam #2 (lam #3 (app #2# #3#)))) (lam #4 #4#))	nil
E1	(app (lam #1 (lam ◊ (app #1# #1#))) (lam #2 (lam #3 (app #2# #3#))))	nil
E2	(lam #1 (lam ◊ (app #1# #1#)))	nil
R2	(clos (lam #1 (lam ◊ (app #1# #1#))) nil)	
E3	(lam #1 (lam #2 (app #1# #2#)))	nil
R3	(clos (lam #1 (lam #2 (app #1# #2#))) nil)	
A4	(clos (lam #1 (lam ◊ (app #1# #1#))) nil)	(clos (lam #2 (lam #3 (app #2# #3#))) nil)
•E5	(lam ◊ (app #1 #1#))	((#1# clos (lam #2 (lam #3 (app #2# #3#))) nil))
R5	(clos (lam ◊ (app #1 #1#)) ((#1# clos (lam #2 (lam #3 (app #2# #3#))) nil)))	
R4	(clos (lam ◊ (app #1 #1#)) ((#1# clos (lam #2 (lam #3 (app #2# #3#))) nil)))	
R1	(clos (lam ◊ (app #1 #1#)) ((#1# clos (lam #2 (lam #3 (app #2# #3#))) nil)))	
E6	(lam #1 #1#)	nil
R6	(clos (lam #1 #1#) nil)	
A7	(clos (lam ◊ (app #1 #1#)) ((#1# clos (lam #2 (lam #3 (app #2# #3#))) nil)))	(clos (lam #4 #4#) nil)
•E8	(app #1 #1#)	((◊ clos (lam #2 #2#) nil) (#1# clos (lam #3 (lam #4 (app #3# #4#))) nil))
E9	#2	((◊ clos (lam #1 #1#) nil) (#2# clos (lam #3 (lam #4 (app #3# #4#))) nil))
R9	(clos (lam #1 (lam #2 (app #1# #2#))) nil)	
E10	#2	((◊ clos (lam #1 #1#) nil) (#2# clos (lam #3 (lam #4 (app #3# #4#))) nil))
R10	(clos (lam #1 (lam #2 (app #1# #2#))) nil)	
A11	#3=(clos (lam #1 (lam #2 (app #1# #2#))) nil)	#3#
•E12	#3=(lam #1 (app #2 #1#))	((#2# clos (lam #2# #3#) nil))
R12	(clos #3=(lam #1 (app #2 #1#)) ((#2# clos (lam #2# #3#) nil)))	
R11	(clos #3=(lam #1 (app #2 #1#)) ((#2# clos (lam #2# #3#) nil)))	
R8	(clos #3=(lam #1 (app #2 #1#)) ((#2# clos (lam #2# #3#) nil)))	
R7	(clos #3=(lam #1 (app #2 #1#)) ((#2# clos (lam #2# #3#) nil)))	
R0	(clos #3=(lam #1 (app #2 #1#)) ((#2# clos (lam #2# #3#) nil)))	◊ = target of unlisted variable pointer

Table B.8 Graph **EV-AP** reduction of ((-2 1 1)(-2 1 0)(-1 0)).

step	term | stack | env	phi
D0	(app (lam _ (app O O)) (lam _ (lam _ (app lam _ (app O (app 1 O))) (app (lam _ (app (app O 1) O)) 1))))) | nil | nil	0
D1	(lam _ (app O O)) | ((susp (lam _ (lam _ (app lam _ (app O (app 1 O))) (app (lam _ (app (app O 1) O)) 1)))) nil)) | nil	0
•D2	(app O O) | nil | ((susp (lam _ (lam _ (app lam _ (app O (app 1 O))) (app (lam _ (app (app O 1) O)) 1)))) nil))	0
D3	O | (#1=(susp (susp (lam _ (lam _ (app lam _ (app O (app 1 O))) (app (lam _ (app (app O 1) O)) 1)))) nil))) | (#1#)	0
D4	#1=(lam _ (lam _ (app lam _ (app O (app 1 O))) (app (lam _ (app (app O 1) O)) 1)))) | ((susp #1# nil)) | nil	0
•D5	#1=(lam _ (app lam _ (app O (app 1 O))) (app (lam _ (app (app O 1) O)) 1)))) | nil | ((susp (lam _ _ #1#) nil))	0
D6	#1=(lam _ (app (app O 1 O)))) (app (lam _ (app (app O 1) O)) 1)) | (\) | (-1 (susp (lam _ (lam _ #1#)) nil))	-1
D7	#1=(lam _ (app O (app 1 O)))) | ((susp #2=(app (lam _ (app app O 1) O)) 1) #3=(-1 (susp (lam _ (lam _ (app (lam _ #1# #2#))) nil))) \) | #3#	-1
•D8	#1=(app O (app 1 O)) | (\) | ((susp #2=(app (lam _ (app app O 1) O)) 1) #3=(-1 (susp (lam _ (lam _ (app (lam _ #1# #2#))) nil))) . #3#)	-1
D9	O | ((susp #1=(app 1 O) #4=((susp #2=(app (lam _ (app app O 1) O)) 1) #3=(-1 (susp (lam _ (lam _ (app (lam _ (app O #1#)) #2#)))) nil))) . #3#) \) | #4#	-1
D10	#1=(app (lam _ (app app O 1) O)) 1) | ((susp #2=(app 1 O) ((susp #1# #3=(-1 (susp (lam _ (lam _ (app lam _ (app O #2#)) #1#))) nil))) . #3#) \) | #3#	-1
D11	#1=(lam _ (app app O 1) O)) | (#4=(susp (lam _ (lam _ (app lam _ (app O #2=(app 1 O))) #3=(app #1# 1)))) nil) (susp #2# ((susp #3# #5=(-1 #4#)) . #5#)) \) | #5#	-1
•D12	#1=(app app O 1) O) | ((susp #2=(app 1 O) ((susp #3=(app (lam _ #1#) 1) #4=(-1 #5=(susp (lam _ (lam _ (app lam _ (app O #2#)) #3#))) nil))) . #4#) \) | (#5# . #4#)	-1
D13	#1=(app O 1) | (#4=(susp (lam _ (lam _ (app lam _ (app O #2=(app 1 O))) #3=(app (lam _ (app #1# O)) 1)))) nil) (susp #2# ((susp #3# #5=(-1 #4#)) . #5#)) \) | (#4# . #5#)	-1
D14	O | (-1 #3=(susp (lam _ (lam _ (app lam _ (app O #1=(app 1 O))) #2=(app (lam _ (app app O 1) O)) 1)))) nil) (susp #1# ((susp #2# #4=(-1 #3#)) . #4#)) \) | (#3# . #4#)	-1
D15	#1=(lam _ (lam _ (app lam _ (app O #2=(app 1 O))) #3=(app (lam _ (app app O 1) O)) 1)))) | (-1 #4=(susp #1# nil) (susp #2# ((susp #3# #5=(-1 #4#)) . #5#)) \) | nil	-1
•D16	#1=(lam _ (app lam _ (app O #2=(app 1 O))) #3=(app (lam _ (app app O 1) O)) 1)))) | (#4=(susp (lam _ _ #1#) nil) (susp #2# ((susp #3# #5=(-1 #4#)) . #5#)) \) | (-1)	-1
•D17	#3=(app (lam _ (app O #1=(app 1 O))) #2=(app (lam _ (app app O 1) O)) 1)) | ((susp #1# ((susp #2# #4=(-1 #5=(susp (lam _ (lam _ #3#)) nil))) . #4#) \) | (#5# -1)	-1
D18	#1=(lam _ (app O #3=(app 1 O))) | ((susp #2=(app (lam _ (app app O 1) O)) 1) #6=(#4=(susp (lam _ (lam _ (app #1# #2#))) nil) -1)) (susp #3# ((susp #2# #5=(-1 #4#)) . #5#)) \) | #6#	-1
•D19	#2=(app O #1=(app 1 O)) | ((susp #1# (susp #3=(app (lam _ (app app O 1) O)) 1) #4=(-1 #5=(susp (lam _ (lam _ (app (lam _ #2#) #3#)))) nil))) . #4#) \) | ((susp #3# #6=(#5# -1)) . #6#)	-1
D20	O | ((susp #1=(app 1 O) #6=((susp #2=(app (lam _ (app app O 1) O)) 1) #3=(#4=(susp (lam _ (lam _ (app lam _ (app O #1#)) #2#))) nil) -1)) . #3#)) (susp #1# ((susp #2# #5=(-1 #4#)) . #5#)) \) | #6#	-1
D21	#1=(app (lam _ (app app O 1) O)) 1) | ((susp #2=(app 1 O) ((susp #1# #3=(#4=(susp (lam _ (lam _ (app (lam _ (app O #2#)) #1#))) nil) -1)) . #3#)) (susp #2# ((susp #1# #5=(-1 #4#)) . #5#)) \) | #3#	-1
D22	#1=(app (lam _ (app app O 1) O)) | (-1 (susp #2=(app 1 O) ((susp #3=(app (lam _ #1# 1) #4=(#5=(susp (lam _ (lam _ (app (lam _ (app O #2#)) #3#)))) nil) -1)) . #4#)) (susp #2# ((susp #3# #6=(-1 #5#)) . #6#)) \) | #4#	-1
•D23	#1=(app app O 1) O) | ((susp #2=(app 1 O) ((susp #3=(app (lam _ #1#) 1) #4=(#5=(susp (lam _ (lam _ (app lam _ (app O #2#)) #3#)))) nil) -1)) . #4#)) (susp #2# ((susp #3# #6=(-1 #5#)) . #6#)) \) | (-1 . #4#)	-1
D24	#1=(app O 1) | (-1 (susp #2=(app 1 O) ((susp #3=(app (lam _ (app #1# O)) 1) #4=(#5=(susp (lam _ (lam _ (app lam _ (app O #2#)) #3#)))) nil) -1)) . #4#)) (susp #2# ((susp #3# #6=(-1 #5#)) . #6#)) \) | (-1 . #4#)	-1
D25	O | (#3=(susp (lam _ (lam _ (app lam _ (app O #1=(app 1 O))) #2=(app (lam _ (app app O 1) O)) 1)))) nil) -1 (susp #1# ((susp #2# #4=(#3# -1)) . #4#)) (susp #1# ((susp #2# #5=(-1 #3#)) . #5#)) \) | (-1 . #4#) *	-1

* The environment portion of the computation graph is not needed from this point on.

step	term | stack | env	phi
U26	O | (#3=(susp (lam _ (lam _ (app lam _ (app O #1=(app 1 O))) #2=(app (lam _ (app app O 1) O)) 1)))) nil) -1 (susp #1# ((susp #2# #4=(#3# -1)) . #4#)) (susp #1# ((susp #2# #5=(-1 #3#)) . #5#)) \)	-1
U27	(app O #3=(susp (lam _ (lam _ (app lam _ (app O #1=(app 1 O))) #2=(app (lam _ (app app O 1) O)) 1)))) nil)) | (-1 (susp #1# ((susp #2# #4=(#3# -1)) . #4#)) (susp #1# ((susp #2# #5=(-1 #3#)) . #5#)) \)	-1
U28	(app (app O #3=(susp (lam _ (lam _ (app lam _ (app O #1=(app 1 O))) #2=(app (lam _ (app app O 1) O)) 1)))) nil)) O) | ((susp #1# ((susp #2# #4=(#3# -1)) . #4#)) (susp #1# ((susp #2# #5=(-1 #3#)) . #5#)) \)	
U29	(app (app (app O #3=(susp (lam _ (lam _ (app lam _ (app O #1=(app 1 O))) #2=(app (lam _ (app app O 1) O)) 1)))) nil)) O) (susp #1# ((susp #2# #4=(#3# -1)) . #4#))) | ((susp #1# ((susp #2# #5=(-1 #3#)) . #5#)) \)	-1
U30	(app (app (app (app O #3=(susp (lam _ (lam _ (app lam _ (app O #1=(app 1 O))) #2=(app (lam _ (app app O 1) O)) 1)))) nil)) O) (susp #1# ((susp #2# #4=(#3# -1)) . #4#))) (susp #1# ((susp #2# #5=(-1 #3#)) . #5#))) | (\)	-1
U31	(lam _ (app (app (app (app O #3=(susp (lam _ (lam _ (app lam _ (app O #1=(app 1 O))) #2=(app (lam _ (app app O 1) O)) 1)))) nil)) O) (susp #1# ((susp #2# #4=(#3# -1)) . #4#))) (susp #1# ((susp #2# #5=(-1 #3#)) . #5#))) | nil	0

Table B.9 HN-HOR trace for the example of section 4.1.

step	term | stack | env	phi
S0	#1=(susp (lam _ (lam _ (app (lam _ (app (lam _ (app 0 (app 0 1))) (lam _ (app (app 2 0) 1)))) (lam _ (app 0 2)))))) nil) | ((0.0 #1#))	
D1	(lam _ (lam _ (app (lam _ (app (lam _ (app 0 (app 0 1))) (lam _ (app (app 2 0) 1)))) (lam _ (app 0 2)))))) | nil | nil	0
D2	(lam _ (app (lam _ (app (lam _ (app 0 (app 0 1))) (lam _ (app (app 2 0) 1)))) (lam _ (app 0 2)))) | (\) | (-1)	-1
D3	(app (lam _ (app (lam _ (app 0 (app 0 1))) (lam _ (app (app 2 0) 1)))) (lam _ (app 0 2))) | (\ \) | (-2 -1)	2
D4	(lam _ (app (lam _ (app 0 (app 0 1))) (lam _ (app (app 2 0) 1)))) | ((susp (lam _ (app 0 2)) #1=(-2 -1) \ \) | #1#	2
•D5	(app (lam _ (app 0 (app 0 1))) (lam _ (app (app 2 0) 1))) | (\ \) | ((susp (lam _ (app 0 2)) #1=(-2 -1) . #1#)	2
D6	(lam _ (app 0 (app 0 1))) | ((susp (lam _ (app 0 2)) #1=(-2 -1) . #1#)) #2=((susp (lam _ (app (app 2 0) 1)) #2=((susp (lam _ (app 0 2)) #1=(-2 -1) . #1#) . #2#)) \ \) | #2#	2
•D7	(app 0 (app 0 1)) | (\ \) | ((susp (lam _ (app (app 2 0) 1)) #2=((susp (lam _ (app 0 2)) #1=(-2 -1) . #1#) . #2#) . #2#)	2
D8	0 | ((susp (app 0 1) #3=((susp (lam _ (app (app 2 0) 1)) #2=((susp (lam _ (app 0 2)) #1=(-2 -1) . #1#) . #2#) . #2#) \ \) | #3#	2
D9	#1=(lam _ (app (app 2 0) 1)) | ((susp (app 0 1) ((susp #1# #3=((susp (lam _ (app 0 2)) #2=(-2 -1) . #2#) . #3#) \ \) | #3#	2
•D10	#1=(app (app 2 0) 1) | (\ \) | ((susp (app 0 1) ((susp (lam _ #1#) #3=((susp (lam _ (app 0 2)) #2=(-2 -1) . #2#) . #3#) . #3#)	2
D11	#1=(app 2 0) | (#2=(susp (lam _ (app 0 2)) #3=(-2 -1) \ \) | ((susp (app 0 1) ((susp (lam _ #1# 1)) #4=(#2# . #3#) . #4#)) . #4#) | #4#	2
D12	2 | (#4=(susp (app 0 1) ((susp (lam _ (app (app 2 0) 1)) #2=(#3=(susp (lam _ (app 0 2)) #1=(-2 -1)) . #1#) . #2#) #3# \ \) | (#4# . #2#)	2
U13	0 | ((susp (app 0 1) ((susp (lam _ (app (app 2 0) 1)) #2=(#3=(susp (lam _ (app 0 2)) #1=(-2 -1) . #1#) . #2#) #3# \ \) |	2
U14	(app 0 (susp (app 0 1) ((susp (lam _ (app (app 2 0) 1)) #2=(#3=(susp (lam _ (app 0 2)) #1=(-2 -1) . #1#) . #2#) (#3# \ \) |	2
U15	(app (app 0 (susp (app 0 1) ((susp (lam _ (app (app 2 0) 1)) #2=(#3=(susp (lam _ (app 0 2)) #1=(-2 -1) . #1#) . #2#) #3# | (\ \) |	2
U16	(lam _ (app (app 0 (susp (app 0 1) ((susp (lam _ (app (app 2 0) 1)) #2=(#3=(susp (lam _ (app 0 2)) #1=(-2 -1) . #1#) . #2#))) #3#) | (\) |	-1
U17	(lam _ (lam _ (app (app 0 (susp (app 0 1) ((susp (lam _ (app (app 2 0) 1)) #2=(#3=(susp (lam _ (app 0 2)) #1=(-2 -1) . #1#) . #2#))) #3#))) | nil |	0
S1	(lam _ (lam _ (app (app 0 #4=(susp (app 0 1) ((susp (lam _ (app (app 2 0) 1)) #2=(#3=(susp (lam _ (app 0 2)) #1=(-2 -1) . #1#) . #2#))) #3#))) | ((-2.0 #4#) (-2.0 #3#))	
D19	(app 0 1) | nil | ((susp (lam _ (app (app 2 0) 1)) #2=((susp (lam _ (app 0 2)) #1=(-2 -1) . #1#) . #2#)	2
D20	0 | (#1=(susp (lam _ (app 0 2)) #2=(-2 -1)) | ((susp (lam _ (app (app 2 0) 1)) #3=(#1# . #2#)) . #3#)	2
D21	(lam _ (app (app 2 0) 1)) | (#1=(susp (lam _ (app 0 2)) #2=(-2 -1)) | (#1# . #2#)	2
•D22	(app (app 2 0) 1) | nil | (#1=(susp (lam _ (app 0 2)) #2=(-2 -1) . #2#)	2
D23	(app 2 0) | (#1=(susp (lam _ (app 0 2)) #2=(-2 -1)) | (#1# #1# . #2#)	2
D24	2 | (#1=(susp (lam _ (app 0 2)) #2=(-2 -1) #1#) | (#1# #1# . #2#)	2
U25	0 | (#1=(susp (lam _ (app 0 2)) (-2 -1)) #1#) |	2
U26	(app 0 #1=(susp (lam _ (app 0 2)) (-2 -1))) | (#1#) |	2
U27	(app (app 0 #1=(susp (lam _ (app 0 2)) (-2 -1))) #1#) | nil |	2
S2	(lam _ (lam _ (app (app 0 (app (app 0 #1=(susp (lam _ (app 0 2)) (-2 -1))) #1#)) #1#))) | ((-2.0 #1#) (-2.0 #1#) (-2.0 #1#))	
D29	(lam _ (app 0 2)) | nil | (-2 -1)	2
D30	(app 0 2) | (\) | (-3 -2 -1)	3
D31	0 | (-1 \) | (-3 -2 -1)	3
U32	0 | (-1 \) |	3
U33	(app 0 2) | (\) |	3
U34	(lam _ (app 0 2)) | nil |	2
S3	(lam _ (lam _ (app (app 0 (app (app 0 #1=(susp (lam _ (app 0 2)) (-2 -1))) #1#)) (lam _ (app 0 2))))) | ((-2.0 #1#) (-2.0 #1#))	
D36	(lam _ (app 0 2)) | nil | (-2 -1)	2
D37	(app 0 2) | (\) | (-3 -2 -1)	3
D38	0 | (-1 \) | (-3 -2 -1)	3
U39	0 | (-1 \) |	3
U40	(app 0 2) | (\) |	3
U41	(lam _ (app 0 2)) | nil |	2
S4	(lam _ (lam _ (app (app 0 (app (app 0 (lam _ (app 0 2))) #1=(susp (lam _ (app 0 2)) (-2 -1)))) (lam _ (app 0 2))))) | ((-2.0 #1#))	
D43	(lam _ (app 0 2)) | nil | (-2 -1)	2
D44	(app 0 2) | (\) | (-3 -2 -1)	3
D45	0 | (-1 \) | (-3 -2 -1)	3
U46	0 | (-1 \) |	3
U47	(app 0 2) | (\) |	3
U48	(lam _ (app 0 2)) | nil |	2
S5	(lam _ (lam _ (app (app 0 (app (app 0 (lam _ (app 0 2))) (lam _ (app 0 2)))) (lam _ (app 0 2))))) | nil |	

Table B.10 BF-HOR trace of tw10, section 4.4, in raw graph form.

Graph

```
(lam _
 (app
  (app
   (app
    (app 0 #3 = (susp (lam _ (lam _ (app (lam _ (app 0 #1 = (app 1 0))) #2 = (app (lam _ (app (app 0 1) 0)) 1)))) nil))
    0)
   (susp #1# ((susp #2# #4 = (#3# -1)) . #4#)))
  (susp #1# ((susp #2# #5 = (-1 #3#)) . #5#))))
```

BTF

```
(-1 0
  ([ (-2 (-1 0 (1 0)) ((-1 0 1 0) 1)) ({ }) ])
  0
  ([ (1 0) ({ ([ ((-1 0 1 0) 1) ({ ([ (-2 (-1 0 (1 0)) ((-1 0 1 0) 1)) ({ }) ]) -1 }) ]) ]) ([ (-2 (-1 0 (1 0)) ((-1 0 1 0) 1))    ({ }) ]) -1 }) ])
  ([ (1 0) ({ ([ ((-1 0 1 0) 1) ({ -1 ([ (-2 (-1 0 (1 0)) ((-1 0 1 0) 1)) ({ }) ]) }) ]) ]) -1 ([ (-2 (-1 0 (1 0)) ((-1 0 1 0) 1    )) ({ }) ]) }) ]))
```

Spine

```
-1 @ ([ (1 0) ({ ([ ((-1 0 1 0) 1) ({ -1 ([ (-2 (-1 0 (1 0)) ((-1 0 1 0) 1)) ({ }) ]) }) ]) ]) -1 ([ (-2 (-1 0 (1 0)) ((-1 0 1 0)    1)) ({ }) ]) }) ]) ])
   @ ([ (1 0) ({ ([ ((-1 0 1 0) 1) ({ ([ (-2 (-1 0 (1 0)) ((-1 0 1 0) 1)) ({ }) ]) -1 }) ]) ]) ([ (-2 (-1 0 (1 0)) ((-1 0 1 0) 1)    ) ({ }) ]) -1 }) ])
   @ 0
   @ ([ (-2 (-1 0 (1 0)) ((-1 0 1 0) 1)) ({ }) ])
  H 0
```

Table B.11 Graph, BTF and Left Spine depiction of the tw7-2 ohnf.

	Graph
Late VA lookup	(lam _ (app (app (app (app 0 (susp 1 #4 = ((susp 1 #3 = ((susp 0 #2 = ((susp 1 #1 = (-1 (susp 0 ((susp (lam _ (lam _ (app (lam _ (app 0 #5 = (app 1 0))) #6 = (app (lam _ (app (app 0 1) 0)) 1)))) nil))))) . #1#)) (susp 1 #2#))) . #3#))) (susp 0 #4#)) (susp #5# ((susp #6# #3#) . #3#))) (susp 5# ((susp #6# #1#) . #1#))))
Early VA lookup	(lam _ (app (app (app (app 0 #3 = (susp (lam _ (lam _ (app (lam _ (app 0 #1 = (app 1 0))) #2 = (app (lam _ (app (app 0 1) 0)) 1)))) nil)) 0) (susp #1# ((susp #2# #4 = (#3# -1)) . #4#))) (susp #1# ((susp #2# #5 = (-1 #3#)) . #5#))))

	BTF
Late VA lookup	(-1 0 [1 { [1 { [0 { [1 { -1 [0 { [(-2 (-1 0 (1 0)) ((-1 0 1 0) 1)) { }] }] }] -1 [0 { [(-2 (-1 0 (1 0)) ((-1 0 1 0) 1)) { }] }] }] [1 { [1 { -1 [0 { [(-2 (-1 0 (1 0)) ((-1 0 1 0) 1)) { }] }] }] -1 [0 { [(-2 (-1 0 (1 0)) ((-1 0 1 0) 1)) { }] }] }] [0 { [1 { -1 [0 { [(-2 (-1 0 (1 0)) ((-1 0 1 0) 1)) { }] }] }] -1 [0 { [(-2 (-1 0 (1 0)) ((-1 0 1 0) 1)) { }] }] }] [1 { [1 { -1 [0 { [(-2 (-1 0 (1 0)) ((-1 0 1 0) 1)) { }] }] }] -1 [0 { [(-2 (-1 0 (1 0)) ((-1 0 1 0) 1)) { }] }] }] [0 { [1 { [0 { [1 { -1 [0 { [(-2 (-1 0 (1 0)) ((-1 0 1 0) 1)) { }] }] }] -1 [0 { [(-2 (-1 0 (1 0)) ((-1 0 1 0) 1)) { }] }] }] [0 { [1 { -1 [0 { [(-2 (-1 0 (1 0)) ((-1 0 1 0) 1)) { }] }] }] -1 [0 { [(-2 (-1 0 (1 0)) ((-1 0 1 0) 1)) { }] }] }] [0 { [1 { -1 [0 { [(-2 (-1 0 (1 0)) ((-1 0 1 0) 1)) { }] }] }] -1 [0 { [(-2 (-1 0 (1 0)) ((-1 0 1 0) 1)) { }] }] }] [(1 0) { [((-1 0 1 0) 1) { [0 { [1 { -1 [0 { [(-2 (-1 0 (1 0)) ((-1 0 1 0) 1)) { }] }] }] -1 [0 { [(-2 (-1 0 (1 0)) ((-1 0 1 0) 1)) { }] }] }] [1 { [1 { -1 [0 { [(-2 (-1 0 (1 0)) ((-1 0 1 0) 1)) { }] }] }] -1 [0 { [(-2 (-1 0 (1 0)) ((-1 0 1 0) 1)) { }] }] }] [0 { [1 { -1 [0 { [(-2 (-1 0 (1 0)) ((-1 0 1 0) 1)) { }] }] }] -1 [0 { [(-2 (-1 0 (1 0)) ((-1 0 1 0) 1)) { }] }] }] [1 { [1 { -1 [0 { [(-2 (-1 0 (1 0)) ((-1 0 1 0) 1)) { }] }] }] -1 [0 { [(-2 (-1 0 (1 0)) ((-1 0 1 0) 1)) { }] }] }] [(1 0) { [((-1 0 1 0) 1) { -1 [0 { [(-2 (-1 0 (1 0)) ((-1 0 1 0) 1)) { }] }] }] }] -1 [0 { [(-2 (-1 0 (1 0)) ((-1 0 1 0) 1)) { }] }] }]
Early VA lookup	(-1 0 [(-2 (-1 0 (1 0)) ((-1 0 1 0) 1)) { }] 0 [(1 0) { [((-1 0 1 0) 1) { [(-2 (-1 0 (1 0)) ((-1 0 1 0) 1)) { }] -1 }] [(-2 (-1 0 (1 0)) ((-1 0 1 0) 1)) { }] -1 }] [(1 0) { [((-1 0 1 0) 1) { -1 [(-2 (-1 0 (1 0)) ((-1 0 1 0) 1)) { }] }] }] -1 [(-2 (-1 0 (1 0)) ((-1 0 1 0) 1)) { }]]

Table B.12 Graph and BTF depictions of the ohnf of tw7-2 with late and early VA lookups.

C. The BTRD Experimentation System[48]

BTRD is a system designed for experimentation with Lambda Calculus reduction. It evolved as a research vehicle of the Head-Order Reduction (HOR) schemes and representation techniques. The eventual goal of this research is to have a complete, correct and efficient Lambda Calculus reduction engine as the basis of a applied reduction system and a software technology environment.

```
BTRD:
An Experimentation Environment for
Reduction in the Lambda Calculus.

Developed by:
©1987-93 Klaus J. Berkling, and
©1991-93 Nikos B. Troullinos.

Version 1.25  October 19, 1993.
                                  [  OK  ]
```

BTRD was originated by Prof. Klaus Berkling at Syracuse University. This author has developed it further by including the abstract algorithms described in this thesis, by organizing all their variations, by providing tracing/instrumentation capabilities and by tidying up the user interface. BTRD in the form of a stand-alone Macintosh application is available upon request from the authors[49].

A lambda expression presented to BTRD can be reduced by choosing appropriate actions from its menus or by issuing equivalent keyboard commands. BTRD has facilities for choosing among numerous weak, head, and strong reduction strategies with and without sharing; it also includes facilities for gathering statistics on runtime behavior and producing detailed traces of the steps taken by the reduction algorithms. Several tests sets are built in, each consisting of a sequence of unreduced–reduced expression pairs. The size of the BTRD application size is approximately 1.8MB and requires a memory partition of at least 3MB. BTRD is subject to continuous development as new algorithms are incorporated.

Highlights

When the BTRD system is launched two windows appear (Figure C.2). The upper one is an editing window meant to receive and edit an input expression. Input to BTRD takes a special form which relies on integers and parentheses to describe any lambda expression. This form is described in section 2.3. Pressing RETURN makes the contents of the edit window the current focus expression and prints its value into the lower window. The lower window

---

[48] This appendix is written in a self-contained style because it also serves as a user's guide.

[49] School of Engineering and Computer Science, Syracuse University, Syracuse, NY 13244.

is at the same time an output buffer and also a full *Lisp* listener. Licensing requirements of *Macintosh Common Lisp* require that the *Lisp* compiler is excised from distributable stand-alone applications, therefore only a *Lisp* interpreter is included in BTRD.

All text windows of BTRD inherit full, Emacs-like, editing capabilities. Parentheses and other bracketing symbols can be matched easily through live blinking. The "Tools" menu includes two items which provide a quick reference to the basic editing commands. The usual "File" menu operations are applicable.

Figure C.1 Dialog for selecting a BTRD reducer and its applicable options.

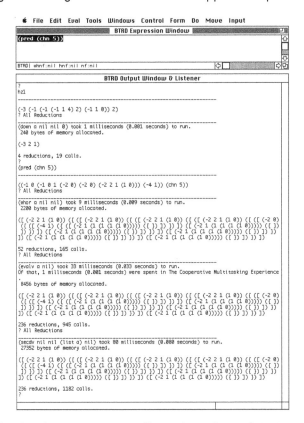

Figure C.2 Input (top) and Output & Listener (bottom) windows of BTRD.

A specific reducer is selected from a dialog which lists all available options (Figure C.1). Strong normal form (*nf*) reducers are listed on the left side of this dialog and weak head normal form (*whnf*) ones on the right side with the exception of the last choice which is one that produces an intermediate head normal form (*hnf*). Sharing and the "Early Variable Argument" lookup tactic can also be chosen, if they are applicable. Currently, 34 variations are permitted. The main simulation/reduction component of BTRD is called *BETAR3* (often abbreviated to *B3*). It is the default simulator when the system starts.

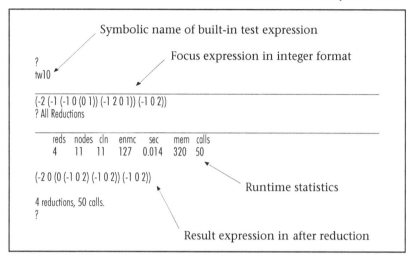

Figure C.3 Explanation of the BTRD output window and listener format.

Field	Description
reds	Number of reductions performed.
nodes	Number of new nodes generated in the process of reduction.
cln	Maximum number of environment cells used.
enmc	Estimated number of memory cycles of a simulated architecture.
sec	Total processing time.
mem	Total memory allocation requested from the *Lisp* system.
calls	Number of iterations of the reduction algorithm.

Table C.1 Runtime statistics fields of the B3 simulator of BTRD.

After a focus expression has been established, it can be reduced to one of the three applicable normal forms. The result and runtime statistical information is displayed in the output window. Figure C.3 shows the reduction of *tw10*, one of the built-in expressions, via the *B3* algorithm. A description of the various fields of the runtime statistics is given in Table C.1. At the end of this Appendix a detailed trace of reduction under *B3* is listed.

In passing we mention that the non-pure extensions of *B3* are not yet frozen and hence are not adequately described in this short guide. Table C.2 demonstrates some of these extensions by repeating the definition of the generalized selection combinator of Table 1.5 (page 20). An equality test for integers, a conditional and a *Y* combinator are utilized.

Coding of a Parameterized Selector
| ```
(-2 Y
   (-2 if (= 0 (+ 2 (chn 1)))
        (-1 Y (-2 if (= 0 (chn 0))
                   2
                      (-1 2 (- 1 (chn 1))))
             (- 1 (chn 1)))
        (-1 2 (- 1 (chn 1))))
   1)
``` |

Table C.2 Definition of a parameterized selector in the syntax of BETAR3.

The BTRD application inherits the memory management system of the underlying development system. When the free space of the *Lisp* heap is exhausted, the cursor turns into the letters "GC" while the system reclaims unused memory. During garbage collection all other activities are suspended. An ephemeral garbage collector which works unobtrusively and mostly in the background is also available. It results in a moderate but continuous performance degradation. It can be enabled by evaluating the form *(egc t)*.
```

Menus

In the next few pages we list all the menus of BTRD. Explanations are kept to a minimum since it is assumed that readers are familiar with the use of an interactive system.

The "File" menu is identical to that of the base development system with the exception of the item "Save Instrum. Data As...". The associated command for this item permits recording in a file of the runtime statistical data produced when one of the test families is run. The output is a tab-delimited text file in which each line holds the data associated with each test.

The "Edit" menu is identical to that of the base development system. It includes the usual commands for cutting, copying, pasting and clearing expressions.

The "Eval" menu is identical to that of the base development system. This menu is only available when the development system is present. In the stand-alone version of BTRD the last four items of this menu are moved to the "Control" menu.

The "Tools" menu is identical to that of the base development system. It includes the ability to query the system about internal symbols, to retrieve documentation strings and to quickly locate the source code of a function by pointing to its name. It also includes commands for activating windows with reference information and dialogs with various environmental settings. In the stand-alone version this menu is simplified.

The "Windows" menu lists all active windows and dialogs. When an item is selected the corresponding window is brought to the front (top) layer.

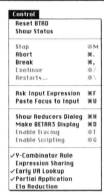

The first two items of the "Control" menu can be used to reset the internal state of the program and to show, in the format of Figure C.5, the values of the registers and of the three areas of the *B3* reducer. *B3* is the more applied, reduction system simulator component of BTRD.

The "Stop" item zeroes the reduction quantum "on the fly". It can be used to stop gracefully very long or infinite reduction sequences.

The "Abort", "Break", "Continue" and "Restarts…" menu items are identical to those of the base development system.

"Ask Input Expression" activates the expression window and selects its current contents so that a focus expression may be readily entered.

"Paste Focus to Input" copies the current focus expression back to the input expression window.

When the next item, "Show Reducers Dialog" is selected, a dialog window appears (shown at left) containing a selection of reduction algorithms. A reducer and its applicable variations can be selected by clicking on the corresponding buttons. To make comparisons between reducers easier, the focus expression is not replaced with its reductum in the case of choosing any reducer other than *B3*.

(*B3*) BTRD can visually display each step of the simulation algorithm *B3* in a special stepper window (Figure C.4). The user can single-step through a reduction while observing all areas of the reducer. Selecting the menu item, "Show BETAR3 Display", brings up this window.

(*B3*) Tracing and scripting is enabled and disabled via the next two menu items. When tracing is enabled, the stepper window is automatically activated when reductions are attempted.

The last five items of the Control menu modify the reduction algorithms as follows:

(B3) "*Y*-Combinator rule" enables the $Yf \rightarrow f(Yf)$ rewrite rule for the Y instead of relying on its lambda definition. When sharing is also selected, a special on-the-fly compilation occurs. The result consists of a cyclic structure with an application node having the operand of the Y as operator and the application node itself as operand.

"Expression Sharing" enables reduction in isolation of closures and subsequent sharing through environment updates.

"Early VA Lookup" enables dereferencing of simple variable arguments (*VA*s) before a closure is formed.

(B3) "Partial Application" enables Curried δ-reductions. When used with expressions containing combinators, reductions do not "fire" until the proper number of arguments are present.

(B3) "Eta Reduction" performs an additional traversal for the purpose of effecting η-reductions.

The "Form" menu includes items which control the format of the output expressions:

When "Application Nodes" is selected a "@" is explicitly shown for each application node. E.g. $Y = (-1 @ (-1 @ 0\ 0)\ (-1 @ 1\ (@\ 0\ 0)))$

"Display as Tree" enables a character-based tree-form display.

"Alphabetic variables" displays expressions using letter variables; they are assigned starting with the identifier "a".

"A - de Bruijn Variables". Binding indices are given as non-negative integers, and lambda sequences as negative integers. This notation is named *BTF* and is described in Chapter 2.

"B - de Bruijn Variables". Binding indices are given as quoted integers, lambdas as indices preceded by a reverse slash (\).

"Raw Internal" is the raw *Lisp* S-expression form. Used for displaying cyclic structures. It takes advantage of the corresponding feature of *Common Lisp*.

"Redisplay" displays again the focus expression in the chosen format.

Via the "Do" menu one can select the operation to be performed on the focus expression:

"One Reduction" does, at most, one reduction. After each reduction the mini-buffer message at the bottom of the expression window is updated to reflect whether the focus expression is in *whnf*, *hnf* or *nf*. Determining this can be costly; therefore a small delay may be noticed with expressions of large size. Currently, only *B3* makes incremental reduction available to the user.

"All Reductions" performs all reductions until either a normal form or the limitations of the system (or of the user) are reached. If the *Control* key is held pressed, printing of the result expression itself is suppressed; only runtime statistics are given in order to circumvent errors arising from the inability of *MCL's* printing mechanism to display expressions of extremely nested structure.

"N Reductions..." does, at most, the specified number of reductions. Currently, only *B3* makes incremental reduction available to the user.

"Beta Extension(s)" performs enough β-extensions to artificially close the focus term in an equivalence preserving manner.

"Eta Extension" performs a single η-extension.

"Expand Sharing" is not currently applicable.

"Compile to Graph" changes the focus expression to a Wadsworth or Aiello–Prini (AP) graph representation.

"Decompile Graph" reverts to BTRD internal raw form from a Wadsworth or AP graph representation.

"Arithmetic Tools" performs RPN-like compilation for expressions with arithmetic subterms.

"Horn-Clause Logic" is currently not applicable.

"Abstract to..." performs combinator abstraction of the focus expression using three different combinator sets and algorithms.

"Explain Symbol..." displays the lambda expression of a predefined symbol in the output expression window.

The "Move" menu always has three choices that vary depending on the actual position within the focus expression. It allows navigation in a equivalence-preserving manner utilizing the nominal machinery of HOR reduction; i.e., unbound variable counts and balancing environment and stack entries.

Finally, the "Input" menu presents several choices for building incrementally well-formed lambda expressions. It is incomplete and not yet user configurable.

Internal Structures

This short presentation of BTRD is concluded with a more detailed look at the internals of its main simulator/reducer component, the *B3*. This portion of the system is designed to be much closer to a realistic reduction-based system in terms of representation and memory management. First, we introduce the various elements and areas of the reducer and we conclude this appendix by providing a complete trace of an expression.

B3 uses three main areas which correspond to the *term*, *stack* and *env* of the abstract *HOR* reducers. The organization of these areas was originated by Berkling and served as a basis to the concrete linear-memory organization of Hilton's *THOR/HORSE* that is discussed in section 8.1. Although the actual allocation is still largely dependent on dynamic *Lisp* structures, a direct mapping to linear-memory is possible.

A streamlined format for the display of the internal state of *B3* has been devised. It is explained in Figure C.5. The contents of these areas together with the values of the internal registers form a complete state. The state can be dumped to the output window on demand or it may be shown continuously via an interactive display (Figure C.4).

The topmost area is a linear depiction of the left spine of the graph from the top to the point of focus. More accurately, it is a depiction of the remaining portion of the left spine after any reductions are performed, and corresponds directly to the stack of the abstract *HOR* reducers. Making the stack grow downwards and placing it above the Problem Graph area is instructive because it provides us with the correct image of descending the left spine by moving the boundary downwards (cf. Table B.11). The Problem Graph is the focus term under reduction and is shown in a straight, left spine format.

The third area is where the deviation from the abstract machines is most evident: two vectors are employed to implement the dynamic, cactus-like environment of Head Order Reduction. In Berkling's current implementation the environment grows downwards from the high index end of two statically-allocated arrays *E1* and *E2* of 10k elements each. The possible entries in these two arrays are shown in Table C.3. *E2* is an auxiliary structure that makes the unavoidable sequential component of a variable lookup more efficient by providing information on directly locating the relevant segment of *E1*.

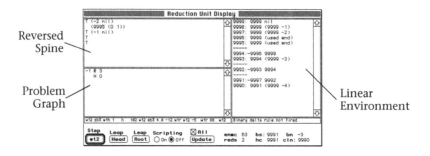

Figure C.4 Reduction Unit Display for the B3/B3-SH simulator/reducer.

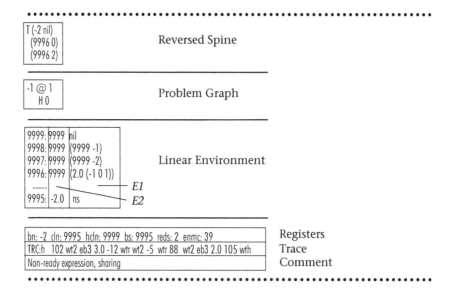

Figure C.5 Areas of the B3/B3-SH simulator/reducer.

	nil	Top of the environment.
	(clnb -n)	An unbound variable pseudo-closure. Default clnb is 9999.
	(-n.0 exp)	An expression, already reduced, tagged with a top ϕ.
E1	(used end)	A potentially shared suspension, currently under reduction.
	(cln exp)	A normal suspension. The expression *exp* is given in BTF.
	cln	The top location of the previous environment segment.
	'ns	The top entry of a new segment.
	int	Current segment top location. The usual entry.
E2	-n.0	Only with a 'ns, the top ϕ of a ready expression.
	-int	Top of current segment; points to top of previous segment.

Table C.3 Possible entries in the B3 environment arrays E1 and E2.

BETAR3 Example Trace

Finally, we present an example of *B3* simulation. This expression under HOR requires only a single down-up traversal; hence it is not demanding enough to illustrate the mechanism of recursively evaluating *hnf* subterms via the stack. More elaborate examples are listed in Appendix B. Before presenting the complete script we show for reference a trace of the same reduction via *HOR-VA* (Table C.4). One can appreciate via this detour the ability of the *BTF* notation to describe the essence of a reduction sequence succinctly.

step	term	stack	env	phi
D0	(-3 (-1 (-1 (-1 1 4) 2) (-1 1 0)) 2)	nil	nil	0
D1	(-2 (-1 (-1 (-1 1 4) 2) (-1 1 0)) 2)	\)	-1)	-1
D2	(-1 (-1 (-1 (-1 1 4) 2) (-1 1 0)) 2)	\\)	-2 -1)	-2
D3	((-1 (-1 (-1 1 4) 2) (-1 1 0)) 2)	\\\)	-3 -2 -1)	-3
D4	(-1 (-1 (-1 1 4) 2) (-1 1 0))	(-1 \\\)	(-3 -2 -1)	-3
•D5	((-1 (-1 1 4) 2) (-1 1 0))	(\\\)	(-1 -3 -2 -1)	-3
D6	(-1 (-1 1 4) 2)	(([(-1 1 0) ({ -1 -3 -2 -1 })]) \\\)	(-1 -3 -2 -1)	-3
•D7	((-1 1 4) 2)	(\\\)	(([(-1 1 0) ({ -1 -3 -2 -1 })]) -1 -3 -2 -1)	-3
D8	(-1 1 4)	(-3 \\\)	(([(-1 1 0) ({ -1 -3 -2 -1 })]) -1 -3 -2 -1)	-3
•D9	(1 4)	(\\\)	(-3 ([(-1 1 0) ({ -1 -3 -2 -1 })]) -1 -3 -2 -1)	-3
D10	1	(-2 \\\)	(-3 ([(-1 1 0) ({ -1 -3 -2 -1 })]) -1 -3 -2 -1)	-3
D11	(-1 1 0)	(-2 \\\)	(-1 -3 -2 -1)	-3
•D12	(1 0)	(\\\)	(-2 -1 -3 -2 -1)	-3
D13	1	(-2 \\\)	(-2 -1 -3 -2 -1)	-3
U14	2	(-2 \\\)		-3
U15	(2 1)	(\\\)		-3
U16	(-1 2 1)	(\ \)		-2
U17	(-2 2 1)	(\)		-1
U18	(-3 2 1)	nil		0

Table C.4 HOR-VA trace for the expression hz1 = (-3 (-1 (-1 (-1 1 4) 2)(-1 1 0)) 2) .

(-3 (-1 (-1 (-1 1 4) 2) (-1 1 0)) 2)
? All Reductions
••

T

-3 @ 2
-1 @ (-1 1 0)
-1 @ 2
-1 @ 4
  H 1

9999: 9999 nil
----
bn: 0 cln: 9999 hcln: 9999 bs: 9999 reds: 0 enmc: 0
TRC: sb3 wth
Start of closure reduction
••

T (-1 nil)

-2 @ 2
-1 @ (-1 1 0)
-1 @ 2
-1 @ 4
  H 1

9999: 9999 nil
9998: 9999 (9999 -1)
----
bn: -1 cln: 9998 hcln: 9999 bs: 9999 reds: 0 enmc: 2
TRC: sb3 wth 16 wth
L' L' / L' top context
••

T (-2 nil)

-1 @ 2
-1 @ (-1 1 0)
-1 @ 2
-1 @ 4
  H 1

9999: 9999 nil
9998: 9999 (9999 -1)
9997: 9999 (9999 -2)
----
bn: -2 cln: 9997 hcln: 9999 bs: 9999 reds: 0 enmc: 4
TRC: sb3 wth 16 wth 16 wth
L' L' / L' top context
••

T (-3 nil)

  @ 2
-1 @ (-1 1 0)
-1 @ 2
-1 @ 4
  H 1

9999: 9999 nil

9998: 9999 (9999 -1)
9997: 9999 (9999 -2)
9996: 9999 (9999 -3)
----
bn: -3 cln: 9996 hcln: 9999 bs: 9999 reds: 0 enmc: 6
TRC: sb3 wth 16 wth 16 wth 16 wth
L' L' / L' top context
••

T (-3 nil)
  (9999 -1)

-1 @ (-1 1 0)
-1 @ 2
-1 @ 4
  H 1

9999: 9999 nil
9998: 9999 (9999 -1)
9997: 9999 (9999 -2)
9996: 9999 (9999 -3)
----
bn: -3 cln: 9996 hcln: 9999 bs: 9999 reds: 0 enmc: 10
TRC: sb3 wth 16 wth 16 wth 16 wth 18 wth
Make closure from variable
••

T (-3 nil)

  @ (-1 1 0)
-1 @ 2
-1 @ 4
  H 1

9999: 9999 nil
9998: 9999 (9999 -1)
9997: 9999 (9999 -2)
9996: 9999 (9999 -3)
9995: 9999 (9999 -1)
----
bn: -3 cln: 9995 hcln: 9999 bs: 9999 reds: 1 enmc: 12
TRC: sb3 wth 16 wth 16 wth 16 wth 18 wth 17 wth
L-AP context
••

T (-3 nil)
  (9995 (-1 1 0))

-1 @ 2
-1 @ 4
  H 1

9999: 9999 nil
9998: 9999 (9999 -1)
9997: 9999 (9999 -2)
9996: 9999 (9999 -3)
9995: 9999 (9999 -1)
----
bn: -3 cln: 9995 hcln: 9999 bs: 9999 reds: 1 enmc: 14
TRC: sb3 wth 16 wth 16 wth 16 wth 18 wth 17 wth 22 wth
AP-AP context

••

T (-3 nil)

  @ 2
-1 @ 4
  H 1

9999: 9999 nil
9998: 9999 (9999 -1)
9997: 9999 (9999 -2)
9996: 9999 (9999 -3)
9995: 9999 (9999 -1)
9994: 9999 (9995 (-1 1 0))
----
bn: -3 cln: 9994 hcln: 9999 bs: 9999 reds: 2 enmc: 16
TRC:sb3 wth 16 wth 16 wth 16 wth 18 wth 17 wth 22 wth 17 wth
L-AP context
••

T (-3 nil)
  (9999 -3)

-1 @ 4
  H 1

9999: 9999 nil
9998: 9999 (9999 -1)
9997: 9999 (9999 -2)
9996: 9999 (9999 -3)
9995: 9999 (9999 -1)
9994: 9999 (9995 (-1 1 0))
----
bn: -3 cln: 9994 hcln: 9999 bs: 9999 reds: 2 enmc: 20
TRC:16 wth 16 wth 16 wth 18 wth 17 wth 22 wth 17 wth 18 wth
Make closure from variable
••

T (-3 nil)

  @ 4
  H 1

9999: 9999 nil
9998: 9999 (9999 -1)
9997: 9999 (9999 -2)
9996: 9999 (9999 -3)
9995: 9999 (9999 -1)
9994: 9999 (9995 (-1 1 0))
9993: 9999 (9999 -3)
----
bn: -3 cln: 9993 hcln: 9999 bs: 9999 reds: 3 enmc: 22
TRC:16 wth 16 wth 18 wth 17 wth 22 wth 17 wth 18 wth 17 wth
L-AP context
••

T (-3 nil)
  (9999 -2)

  H 1

9999: 9999 nil
9998: 9999 (9999 -1)
9997: 9999 (9999 -2)
9996: 9999 (9999 -3)
9995: 9999 (9999 -1)
9994: 9999 (9995 (-1 1 0))
9993: 9999 (9999 -3)
----
bn: -3 cln: 9993 hcln: 9999 bs: 9999 reds: 3 enmc: 26
TRC:16 wth 18 wth 17 wth 22 wth 17 wth 18 wth 17 wth 18 wth
Make closure from variable
••

T (-3 nil)
  (9999 -2)

  H 1

9999: 9999 nil
9998: 9999 (9999 -1)
9997: 9999 (9999 -2)
9996: 9999 (9999 -3)
9995: 9999 (9999 -1)
9994: 9999 (9995 (-1 1 0))
9993: 9999 (9999 -3)
----
bn: -3 cln: 9993 hcln: 9999 bs: 9999 reds: 3 enmc: 29
TRC:18 wth 17 wth 22 wth 17 wth 18 wth 17 wth 18 wth 1 h
Head variable
••

T (-3 nil)
  (9999 -2)

-1 @ 0
  H 1

9999: 9999 nil
9998: 9999 (9999 -1)
9997: 9999 (9999 -2)
9996: 9999 (9999 -3)
9995: 9999 (9999 -1)
9994: 9999 (9995 (-1 1 0))
9993: 9999 (9999 -3)

9992:-9995 ns
----
bn: -3 cln: 9992 hcln: 9999 bs: 9992 reds: 3 enmc: 30
TRC:17 wth 22 wth 17 wth 18 wth 17 wth 18 wth 1 h 106 wth
Non-ready expression, no sharing
••

T (-3 nil)

  @ 0
  H 1

9999: 9999 nil
9998: 9999 (9999 -1)
9997: 9999 (9999 -2)
9996: 9999 (9999 -3)

9995: 9999 (9999 -1)
9994: 9999 (9995 (-1 1 0))
9993: 9999 (9999 -3)

9992:-9995 ns
9991: 9992 (9999 -2)
----
bn: -3 cln: 9991 hcln: 9999 bs: 9992 reds: 4 enmc: 32
TRC:22 wth 17 wth 18 wth 17 wth 18 wth 1 h 106 wth 17 wth
L-AP context
● ●

T (-3 nil)
  (9999 -2)

  H 1

9999: 9999 nil
9998: 9999 (9999 -1)
9997: 9999 (9999 -2)
9996: 9999 (9999 -3)
9995: 9999 (9999 -1)
9994: 9999 (9995 (-1 1 0))
9993: 9999 (9999 -3)

9992:-9995 ns
9991: 9992 (9999 -2)
----
bn: -3 cln: 9991 hcln: 9999 bs: 9992 reds: 4 enmc: 36
TRC:17 wth 18 wth 17 wth 18 wth 1 h 106 wth 17 wth 18 wth
Make closure from variable
● ●

T (-3 nil)
  (9999 -2)

  H 1

9999: 9999 nil
9998: 9999 (9999 -1)
9997: 9999 (9999 -2)
9996: 9999 (9999 -3)
9995: 9999 (9999 -1)
9994: 9999 (9995 (-1 1 0))
9993: 9999 (9999 -3)

9992:-9995 ns
9991: 9992 (9999 -2)
----
bn: -3 cln: 9991 hcln: 9999 bs: 9992 reds: 4 enmc: 41
TRC:18 wth 17 wth 18 wth 1 h 106 wth 17 wth 18 wth 1 h
Head variable
● ●

T (-3 nil)
  (9999 -2)

  H 2

9999: 9999 nil
9998: 9999 (9999 -1)

9997: 9999 (9999 -2)
9996: 9999 (9999 -3)
9995: 9999 (9999 -1)
9994: 9999 (9995 (-1 1 0))
9993: 9999 (9999 -3)

9992:-9995 ns
9991: 9992 (9999 -2)
----
bn: -3 cln: 9991 hcln: 9999 bs: 9992 reds: 4 enmc: 41
TRC:17 wth 18 wth 1 h 106 wth 17 wth 18 wth 1 h 102 wt2
Bound variable head
● ●

T (-3 nil)

  @ 1
  H 2

9999: 9999 nil
9998: 9999 (9999 -1)
9997: 9999 (9999 -2)
9996: 9999 (9999 -3)
9995: 9999 (9999 -1)
9994: 9999 (9995 (-1 1 0))
9993: 9999 (9999 -3)

9992:-9995 ns
9991: 9992 (9999 -2)
----
bn: -3 cln: 9991 hcln: 9999 bs: 9992 reds: 4 enmc: 43
TRC:wth 1 h 106 wth 17 wth 18 wth 1 h 102 wt2 -6 wtr wt2
Bound var-arg short
● ●

T

-3 @ 1
  H 2

9999: 9999 nil
9998: 9999 (9999 -1)
9997: 9999 (9999 -2)
9996: 9999 (9999 -3)
9995: 9999 (9999 -1)
9994: 9999 (9995 (-1 1 0))
9993: 9999 (9999 -3)

9992:-9995 ns
9991: 9992 (9999 -2)
----
bn: 0 cln: 9991 hcln: 9999 bs: 9992 reds: 4 enmc: 46
TRC:wth 17 wth 18 wth 1 h 102 wt2 -6 wtr wt2 -5 wtr 88 wt2
Binary delta rule not fired
● ●

reds	nodes	cln	enmc	sec	mem	calls
4	2	8	46	0.008	120	19

(-3 2 1)
4 reductions, 19 calls.

D. Test Expressions and Runtime Statistics

The general set of test expressions *all** is listed below (Table D.1). Quoted integers denote equivalent Church Numeral expressions. Test cases are given as pairs; result *wfes* have names identical to those of problem *wfes* but starting with "r". Measurements for this and

tqq	(-7 (-4 7 4 (-1 6 5))(-4 5 1)(-2 3 4) 2 3)	rqq	(-7 3 0 (-1 2 1))
tww	(-7 (-4 7 4 (-1 -4 3 1 (2 9 11)) 5))(-4 5 1)(-2 3 4) 2 3)	rww	(-7 3 0 (-4 4 1 (2 4 6)))
tqw	(-7 (-4 7 (-3 (-4 7 4 (-1 (-4 3 1 (2 9 11)) 5))(-4 5 1) (-2 3 4) 2 3) 4 (-1 6 5))(-4 5 1)(-2 3 4) 2 3)	rqw	(-7 3 (-3 6 0 (-4 4 1 (2 4 6))) 0 (-1 2 1))
tt1	(-4 (-3 2 1 0) 3 2)	rt1	(-5 4 3 0)
tt2	((-2 1 (1 (1 (1 0))))(-2 1 (1 (1 0))))	rt2	('81)
tt21	((-2 1 (1 0))(-2 1 0))	rt21	(-2 1 0)
tt22	((-2 1 (1 0))(-2 1 (1 0)))	rt22	('4)
tt23	((-2 1 (1 0))(-2 1 (1 (1 0))))	rt23	('9)
tt32	((-2 1 (1 (1 0)))(-2 1 (1 0)))	rt32	('8)
tt3	((-2 1 0)(-2 1 0))	rt3	(-2 1 0)
tt4	((-2 1)(-2 0))	rt4	(-3 0)
tt5	(-4 3)	rt5	(-4 3)
tt6	((-3 2 0 (1 0))(-2 1)(-2 1))	rt6	(-1 0)
tt7	((-4 3 1 (2 1 0))('5)('7))	rt7	('1))
tt8	((-3 2 (1 0))('5)('7))	rt8	('3)
tt9	((-2 0 1)('2)('7))	rt9	('18)
tt10	((-4 3 1 (2 1 0))('5)('5))	rt10	('1))
tt11	((-3 2 (1 0))('5)('5))	rt11	('2)
tt12	((-2 0 1)('4)('3))	rt12	('6)
tt13	((-2 0 1)('3)('4))	rt13	('8)
tt14	(-3 (-3 2 1 0 1 4)(-3 2 5 3 1 0))	rt14	(-5 1 4 2 0 1 3)
tw1	(-3 (-1 0 3)(1 2)((-1 0 3)(1 2) 0 (-1 0 0)))	rw1	(-3 1 2 2 (1 2 2 0 (-1 0 0)))
tw2	(-2 (-1 1 0 (-1 0 3))((-1 1 0 (-1 0 3))(-1 0 2)))	rw2	(-2 0 (0 (-1 0 2)(-1 0 2))(-1 0 2))
tw3	(-2 (-1 0 (0 (0 2)))((-2 0 1) 0))	rw3	(-2 1 0 0 0)
tw4	(-2 (-1 0 2 0)(-1 0 ((-1 0) 1)))	rw4	(-2 1 0 (-1 0 1))
tw5	((-1 0 0)((-1 (-1 0) 0)))	rw5	(-1 0)
tw6	(-3 (-1 0 (3 0))((-1 0 2) 0))	rw6	(-3 0 1 (2 (0 1)))
tw7	(-2 (-1 0 (1 0))((-1 0 1 0) 1))	rw7	(-2 1 0 1 (0 (1 0 1)))
tw8	(-1 (-1 (-1 (-1 0 0 2) 0)(0 1))(-1 0))	rw8	(-1 0 0 (-1 0))
tw9	(-2 (-1 2 0 (1 0))(-1 0 1))	rw9	(-2 1 (-1 0 1)(0 (-1 0 1))
tw10	(-2 (-1 (-1 0 (0 1))(-1 2 0 1))(-1 0 2))	rw10	(-2 0 (0 (-1 0 2)(-1 0 2))(-1 0 2))
tw11	(-2 (-1 (-1 0 (0 1))(-2 3 1 2))(-1 0 2))	rw11	(-3 1 (-1 2 (-1 0 4)(-1 0 4))(-1 0 3))
tw12	(-2 (-1 (-1 0 (0 1))(-2 2 0 1))(-1 0 2))	rw12	(-3 0 2 (-1 0 3 (-1 0 4)))
hz1	(-3 (-1 (-1 (-1 1 4) 2)(-1 1 0)) 2)	rz1	(-3 2 1)
lev1	(-2 (-1 0 2 (0 1))(-1 (-1 0) 0))	rev1	(-2 1 0)
lev2	((-3 2 1 (2 0))(-1 (-1 0) 0))	rev2	(-2 1 0)
tmh4	(-3 (-1 0 3)(1 2)((-1 0 3)(1 2) 0 (-1 0))	rmh4	(-3 1 2 2 (1 2 2 0 (-1 0))
tmh9	((-2 0 1 1 1 1)((-2 0 1 1)(-1 0)(-1 0)))	rmh9	(-1 0 (-1 0)(-1 0)(-1 0)(-1 0)) continued on the next page

Table D.1 The general-purpose test expression set *all**.

other test sets were presented in Chapter 7. Normal order strategies perform a total of 13 147 reductions to produce the *nf*'s of all the expressions of this test set. Table D.2 shows run-time statistics under *HOR-SH-VA*.

tw7-2	((-2 (-1 0 (1 0))((-1 0 1 0) 1))(-2 (-1 0 (1 0)) ((-1 0 1 0) 1)))	rw7-2	(-1 0 (-2 1 0 1 (0 (1 0 1))) 0 (-1 1 (-2 1 0 1 (0 (1 0 1))) 1 0 (1 (-2 1 0 1 (0 (1 0 1))) 1)(0 (1 (-2 1 0 1 (0 (1 0 1))) 1 0 (1 (-2 1 0 1 (0 (1 0 1))) 1))))(0 (0 (-2 1 0 1 (0 (1 0 1))) 0 (-1 1 (-2 1 0 1 (0 (1 0 1))) 1 0 (1 (-2 1 0 1 (0 (1 0 1))) 1) (0 (1 (-2 1 0 1 (0 (1 0 1))) 1 0 (1 (-2 1 0 1 (0 (1 0 1))) 1)))))))
tw10-2	((-2 (-1 (-1 0 (0 1))(-1 2 0 1))(-1 0 2))(-2 (-1 (-1 0 (0 1)) (-1 2 0 1))(-1 0 2)))	rw102	(-1 0 (0 (-1 0 (-2 0 (0 (-1 0 2)(-1 0 2))(-1 0 2))) (-1 0 (-2 0 (0 (-1 0 2)(-1 0 2))(-1 0 2)))(-1 0 (-2 0 (0 (-1 0 2)(-1 0 2))(-1 0 2))))
nbot	((-1 (-5 (-8 0) 4 2 1 4 0 2 3) 0 0 0 0)((-1 0 0)(-1 0 0)))	rbot	(-1 0)
pred7	((-1 0 (-1 0 1 (-2 0)(-2 0)(-2 2 1 (1 0))(-4 1))('7))	rred7	('6)
pred9	((-1 0 (-1 0 1 (-2 0)(-2 0)(-2 2 1 (1 0))(-4 1))('9))	rred9	('8)
fsa2	((-1 0 0)(-1 (-1 0 0)(-1 1 (0 (-1 0)))))	rsa2	(-1 0)
fs2n8	((-1 0 0)((-1 0 0)((-1 0 0)((-1 0 0)((-1 0 0)(-1 0 0) ((-1 0 0)((-1 0 0)(-2 0)))))))))	rs2n8	(-1 0)
fs4n8	(-1 (-1 0 1 (0 0))(-1 (-1 0 1 (0 0))(-1 (-1 0 1 (0 0)) (-1 (-1 0 1 (0 0))(-1 (-1 0 1 (0 0))(-1 (-1 0 1 (0 0)) (-1 (-1 0 1 (0 0))(-1 (-1 0 1 (0 0))(-2 0)))))))))	rs4n8	(-2 0)
t3s0	((-2 1 (1 0))(-2 1 (1 0))(-2 1 (1 0))(-3 1 (2 1 0))('0))	r3s0	(-2 1 (1 (1 (1 (1 (1 (1 (1 (1 (1 (1 (1 (1 (1 (1 (1 0))))))))))))))))
pones	(-1 (-2 1 (-1 0)(1 (-1 2)))(-1 (-1 0 0 (0 0))(0 1)))	rones	(-2 1 1 (1 1)(1 1 (1 1))(1 1 (1 1))(1 1 (1 1))) (1 1 (1 1)(1 1 (1 1))(1 1 (1 1)(1 1 (1 1)))) (1 1 (1 1) (1 1 (1 1))(1 1 (1 1))(1 1 (1 1))) (1 1 (1 1) (1 1 (1 1))(1 1 (1 1))(1 1 (1 1)))))))
ashar1	((-1 (-3 2 (1 0)) 0 0)((-4 3 1 (2 1 0)) ((-3 2 (1 0))('5)('4))('7)))	rshar1	('729)
ashar2	((-1 (-3 2 (1 0)) 0 (-3 2 (1 0)) 0 0)((-4 3 1 (2 1 0)) ((-3 2 (1 0))('3)('2)('4)))	rshar2	('1000)
ashar3	((-1 (-3 2 (1 0)) 0 (-3 2 (1 0)) 0 ((-3 2 (1 0)) 0 0)) ((-4 3 1 (2 1 0))((-3 2 (1 0))('2)('1)('2)))	rshar3	('256)
ashar4	((-1 ((-3 2 (1 0))((-3 2 (1 0)) 0 0)((-3 2 (1 0)) 0 0)) ((-4 3 1 (2 1 0))((-3 2 (1 0))('2)('1)('3)))	rshar4	('625)
lev3	((-3 (-1 3 (0 1))(-1 0)((-1 3 (0 1))(-1 2)))(-1 (-1 0) (0 0)))	rev3	(-2 0 0 (1 1))
wads34	(-2 ((-1 0 2 0)(-1 0 ((-1 0) 1))))	rads34	(-2 1 0 (-1 0 1))
wads45	(-2 (-1 0 (0 2))((-2 0 1) 0))	rads45	(-2 1 0 0)

Table D.1 (Continued) The general-purpose test expression set all*.

Test	Reductions	Calls down	Calls up	Time†	Memory	Cost	Reds/sec†
tqq	4	21	13	0.002	728	1.21	2000
tww	5	30	20	0.003	1048	1.77	1667
tqw	9	48	30	0.003	1592	2.19	3000
tt1	2	12	8	0.001	408	0.64	2000
tt2	14	294	253	0.019	9648	13.54	737
tt21	4	22	11	0.002	552	1.05	2000
tt22	5	35	21	0.003	944	1.68	1667
tt23	6	52	35	0.003	1480	2.11	2000
tt32	8	61	38	0.004	1664	2.58	2000
tt3	2	13	8	0.001	384	0.62	2000
tt4	1	10	7	0.001	360	0.60	1000
tt5	0	5	5	0.001	256	0.51	0
tt6	4	14	5	0.001	344	0.59	4000
tt7	6	68	54	0.004	2152	2.93	1500
tt8	9	140	115	0.010	4472	6.69	900
tt9	22	602	538	0.042	20312	29.21	524
tt10	6	60	46	0.004	1864	2.73	1500
tt11	9	112	87	0.008	3464	5.26	1125
tt12	14	230	190	0.015	7336	10.49	933
tt13	16	309	263	0.022	10064	14.88	727
tt14	4	28	19	0.003	896	1.64	1333
tw1	2	25	19	0.002	784	1.25	1000
tw2	2	21	17	0.002	728	1.21	1000
tw3	5	29	14	0.002	656	1.15	2500
tw4	3	16	9	0.001	424	0.65	3000
tw5	3	10	3	0.001	160	0.40	3000
tw6	2	15	10	0.002	448	0.95	1000
tw7	2	16	11	0.002	448	0.95	1000
tw8	4	17	7	0.001	336	0.58	4000
tw9	1	15	13	0.002	528	1.03	500
tw10	4	34	24	0.002	1040	1.44	2000
tw11	4	37	27	0.003	1208	1.90	1333
tw12	5	31	18	0.002	880	1.33	2500
hz1	4	17	8	0.001	432	0.66	4000
lev1	4	17	7	0.001	392	0.63	4000
lev2	4	17	7	0.001	392	0.63	4000
tmh4	2	24	18	0.002	752	1.23	1000
tmh9	17	68	26	0.004	1280	2.26	4250
tw7-2	8	170	148	0.012	5656	8.24	667
tw10-2	13	137	105	0.009	4544	6.39	1444
nbot	13	28	2	0.001	320	0.57	13000
pred7	80	463	250	0.027	11912	17.93	2963
pred9	120	718	393	0.041	18784	27.75	2927
fsa2	12	61	27	0.004	1224	2.21	3000
fs2n8	16	52	12	0.003	648	1.39	5333
fs4n8	32	131	43	0.009	3248	5.41	3556
t3s0	34	259	168	0.016	7304	10.81	2125
pones	22	219	154	0.015	6136	9.59	1467
ashar1	47	1792	1660	0.124	61440	87.28	379
ashar2	42	2606	2491	0.188	91296	131.01	223
ashar3	37	982	884	0.069	33296	47.93	536
ashar4	60	1635	1468	0.113	55040	78.86	531
lev3	8	41	18	0.002	904	1.34	4000
wads34	3	16	9	0.002	424	0.92	1500
wads45	4	21	10	0.002	488	0.99	2000
Totals	769	11906	9846	0.820	383520	559.78	2192
† Macintosh Common Lisp; 25MHz 68030 processor; no instrumentation.							

Table D.2 HOR-SH-VA reduction of the all* test set.

Bibliography

[Abadi 90]   Abadi, M., Cardelli, L., Curien, P.-L. and Lévy, J.-J Explicit Substitutions.
                  In *Proc. seventeenth ACM Symposium on Principles of Programming
                  Languages,* ACM, ACM Press, 1990, pp. 1-16.

[Abadi 91]   Abadi, M. Explicit Substitutions. *Journal of Functional Programming 1*, L.
                  Cardelli (1991), 375-416.

[Aiello 81]   Aiello, Luigia and Prini, Gianfranco An Efficient Interpreter for the
                  Lambda-Calculus. *Journal of Computer and System Sciences 23*(1981), 383–
                  424.

[Backus 90]   Backus, J., Williams, J. and Wimmers, E. An Introduction to the
                  Functional Language FL. In *Research Topics in Functional Programming*.
                  Addison-Welsey, Turner, D.A., pp. 219-247, Reading, MA, 1990.

[Barendregt 84]   Barendregt, H.P. *The Lambda Calculus, Its Syntax and Semantics, Revised
                  Edition,* North-Holland, Amsterdam , Studies in Logic(1984).

[Barendregt 86]   Barendregt, H.P., Kennaway, J.R., Klop, J.W. and Sleep, M.R., "Needed
                  Reduction and Spine Strategies for the Lambda Calculus", Centrum voor
                  Wiskunde en Informatica, Report, no. CS-R8621, Amsterdam, 1986.

[Berkling 77]   Berkling, K.J. Reduction Languages for Reduction Machines. In *Future
                  Systems,* Infotech International Limited, Nicholson House, Maidenhead
                  Berkshire, England, 1977, pp. 79-116.

[Berkling 82]   Berkling, K.J. A Consistent Extension of the Lambda-Calculus As a Base
                  for Functional Programming Languages. *Information and Control
                  55*(1982), 89-101.

[Berkling 87]   Berkling, K.J. Head Order Reduction: a Graph Reduction Scheme for the
                  Operational Lambda Calculus. In *Graph Reduction: Proceedings of a
                  Workshop, Santa Fé, New Mexico, USA,* Fasel, J.H. and Keller, R.M.,
                  Springer-Verlag, New York, NY, 1987, pp. 26-48.

[Berkling 92]   Berkling, Klaus J., *Computer for Reducing Lambda Calculus Expressions
                  Employing Variable Containing Applicative Language Code,* US Patent No.
                  5,99,450, March, 24 1992, .

[Bruijn 72]   de Bruijn, N.G. A Lambda Calculus Notation with Nameless Dummies, a
                  Tool for Automatic Formula Manipulation, with Application to the
                  Church-Rosser Theorem. *Indagationes Mathematicae 34*(1972), 381-392.

[Burge 75]   Burge, W.H. *Recursive Programming Techniques,* Addison-Wesley, Reading,
                  MA (1975).

[Cousineau 85]   Cousineau, G., Curien, P.L. and Mauny, M. The Categorical Abstract Machine. In *Functional Programming Languages and Computer Architecture*. Springer-Verlag, Jouannaud, J.-P., pp. 50-64, Berlin, 1985.

[Curien 86]   Curien, P.-L. *Categorical Combinators, Sequential Algorithms and Functional Programming*, John Wiley, New York (1986).

[Curry 58]   Curry, H.B., Craig, W. and Feys, R. *Combinatory Logic, Volume 1*, North-Holland, Amsterdam (1958).

[Field 88]   Field, A.J. and Harrison, P.G. *Functional Programming*, Addison-Wesley, Wokingham, Berkshire (1988).

[Field 90]   Field, John H. On Laziness and Optimality in Lambda Interpeters: Tools for Specification and Analysis. In *Seventeenth ACP Symposium on Principles of Programming Languages*, ACM, ACM Press, 1990, pp. 1-15.

[Field 91]   Field, John H. *Incremental Reduction in the Lambda Calculus and Related Substitution Systems*, Ph.D. dissertation, Cornell University, Ithaca, NY, November 1991.

[Frandsen 91]   Frandsen, G.S. and Sturtivant, C., "What is an Efficient Implementation of the Lambda-Calculus?", Computer Science Department, Aarhus University, no. DAIMI-344, Aarhus, Also in LNCS 523, pp 289-312, 1991.

[Greene 85]   Greene, Kevin J. *A Fully Lazy Higher Order Purely Functional Language with Reduction Semantics*, Ph.D. dissertation, Syracuse University, Syracuse, NY, December 1985.

[Henderson 80]   Henderson, P. *Functional Programming: Application and Implementation*, Prentice-Hall, London (1980).

[Henderson 82]   Henderson, P., Jones, G.A. and Jones, S.B., "The Lispkit Manual", Oxford University Computing Laboratory, Technical monograph, no. PRG-2, 1982.

[Hilton 90a]   Hilton, Michael L. and Berkling, Klaus J., "User's Manual and Programming Guide for THOR/HORSE (Version 1.0)", Syracuse University, CASE Center Techical Report, no. 9006, Syracuse, NY, May 1990.

[Hilton 90b]   Hilton, Michael Lee, "Implementation of Declarative Languages", CASE Center, Syracuse University, Ph.D. Dissertation, no. 9008, Syracuse, New York, June 1990.

[Hilton 90c]   Hilton, Michael L. and Berkling, Klaus J., "Modifying the von Neumann Architecture to Directly Support an Alternate Theory of Computation," September 1990, Draft.

[Hilton 91]   Hilton, Michael L. and Berkling, Klaus J., "A Semantics-Directed Architecture for Reducing the Lambda Calculus," September 1991, Draft.

[Hindley 86]   Hindley, J.R. and Seldin, J.P. *Introduction to Combinators and Lambda-Calculus,* Cambridge University Press, Cambridge (1986).

[Hoare 85]   Hoare, C.A.R. *Communicating Sequential Processes,* Prentice-Hall International (1985).

[Hughes 90]	Hughes, R.J.M. Why Functional Programming Matters. In *Research Topics in Functional Programming*. Addison-Welsey, Turner, D.A., pp. 17-42, Reading, MA, 1990.
[Kathail 90]	Kathail, Vinod K. *Optimal Interpreters for Lambda-Calculus Based Functional Languages*, Ph.D. dissertation, Massachusetts Institute of Technology, May 1990.
[Kennaway 88]	Kennaway, J.R. and Sleep, M.R. Director Strings As Combinators. *ACM Transactions on Programming Languages and Systems 10*, 4 (1988), 602-626.
[Kieburtz 86]	Kieburtz, Richard B. When Chasing Your Tail Saves Time. *Information Processing Letters 23*(December 1986), 321–324.
[Kluge 93]	Kluge, Werner E., "A User's Guide for the Reduction System Pi-Red," January 1993, University of Kiel, Draft.
[Lamping 90]	Lamping, John An Algorithm for Optimal Lambda Calculus Reduction. In *Seventeenth Annual ACM Symposium on Principles of Programming Languages,* ACM, ACM Press, 1990, pp. 16-30.
[Landin 64]	Landin, P.J. The Mechanical Evaluation of Expressions. *The Computer Journal 6*, 4 (January 1964), 308-320.
[Landin 65]	Landin, P.J. A Correspondence Between ALGOL60 and Church's Lambda-Notation. *Communications of the ACM 8*(1965), 89-101, 158-165.
[Langendoen 93]	Langendoen, Koen *Graph Reduction on Shared Memory Multiprocessors*, Ph.D. dissertation, University of Amsterdam, April 1993.
[Lévy 88]	Lévy, J-J. Sharing in the Evaluation of Lambda Expressions. In *Programming of Future Generation Computers II*. North-Holland, Fuchi, K. and Kott, L., 1988.
[McCarthy 60]	McCarthy, J. Recursive Functions of Symbolic Expressions. *Communications of the ACM 3*(1960), 184-195.
[McCarthy 65]	McCarthy, J., Abrahams, P.W., Edwards, D.J., Hart, T.P. and Levin, M.E. *Lisp 1.5 Programmers' Manual,* MIT Press, Cambridge, MA (1965).
[Milner 80]	Milner, R. *A Calculus of Communicating Systems,* Springer-Verlag, Berlin (1980).
[Oberhauser 87]	Oberhauser, H.-G. On the Correspondence of Lambda Style Reduction and Combinator Style Reduction. In *Graph Reduction: Proceedings of a Workshop at Santa Fé, New Mexico,* Fasel, J.H. and Keller, R.M., Springer-Verlag, New York, NY, 1987, pp. 1-25.
[PeytonJones 87]	PeytonJones, S.L. *The Implementation of Functional Programming Languages,* Prentice Hall, London (1987).
[Revesz 88]	Revesz, G. *Lambda-Calculus, Combinators, and Functional Programming,* Cambridge University Press, Cambridge (1988).
[Roylance 88]	Roylance, Gerald Expressing Mathematical Subroutines Constructively. In *1988 ACM Conference on Lisp and Functional Programming,* ACM, Snowbird, UT, July 1988, pp. 8–13.

[Schmittgen 92]   Schmittgen, C., Blödorn, H. and Kluge, W. Pi-Red-Star – A Graph Reducer
                   for a Full-Fledged Lambda Calculus. *New Generation Computing 10*, 2
                   (1992), 173-195.

[Staples 80]   Staples, J. Optimal Evaluation of Graph-Like Expressions. *Theoretical
                   Computer Science 10*(1980), 297-310.

[Turner 79]   Turner, D.A. A New Implementation Technique for Applicative
                   Languages. *Software - Practice and Experience 9*(1979), 31-49.

[Wadsworth 71]   Wadsworth, C. P. *Semantics and Pragmatics of the Lambda Calculus*, Ph.D.
                   dissertation, Oxford University, 1971.

[Zhang 89]   Zhang, Sining and Berkling, Klaus, "The Soundness and Completeness of
                   Head Order Reduction", CASE Center, Syracuse University, no. 8907, May
                   1989.

Index